Hydrogen: Energy Vector of the Future

Published by Graham & Trotman

Conference held on 24/25 March 1981

Organized by
Haus der Technik eV, Essen

in conjunction with the
Gesellschaft Deutscher Chemiker eV (GDCh)
Frankfurt/Main

under the direction of Dr-Ing G Beghi
Commission of the European Communities
Ispra Joint Research Centre
Ispra/Varese, Italy

Translation from German by Cave Translations Ltd, London arranged by the Commission of the European Communities, Directorate-General, Information Market and Innovation

Published in English in 1983 by
Graham & Trotman Limited
Sterling House, 66 Wilton Road, London SW1V 1DE
Telephone: 01-821 1123. Telex: 298878 GRAMCO G. Cables: Infobooks London

ISBN: 0 86010 451 6

Originally published in German as *Wasserstoff — Energieträger der Zukunft*

ISBN: 3-8027-0448-7

Reprinted by Antony Rowe Ltd, Chippenham

Contents

HYDROGEN -
ENERGY VECTOR OF THE FUTURE

Conference held on 24/25 March 1981
at Haus der Technik Essen

FOREWORD

by G. Beghi[1]

In the distant future, hydrogen could play a significant role in solving our energy problems.

It has been assuming ever-growing importance in recent years as an energy vector.

Particular attention has been paid to the possible uses of hydrogen as a means of energy storage and energy transport and as a versatile fuel.

Work on hydrogen technology is in progress in a number of European countries, the USA and Japan from various aspects and in various contexts.

The following papers provide a general review of the hydrogen technology of today and tomorrow.

1. Dr-Ing. G. Beghi, Commission of the European Communities, Ispra Joint Research Centre, Ispra/Varese Italy

An introduction to hydrogen technology

by G. Beghi[1]

Why hydrogen?

During recent years there has been increasingly frequent discussion of the utilization of hydrogen. Why this great interest in hydrogen?

Hydrogen is already an important and long-established raw material for the chemical industry today, but the chief reason for this upsurge of interest is the possibility of its use in the future as an energy vector in order to satisfy the growing demands of our society.

Hydrogen has an important part to play in the manufacture of synthetic fuels, as a substitute for ever more costly petroleum products.

With hydrogen as a future means of energy storage, nuclear energy and new energy sources such as solar energy could be better utilized. The storage and transport of hydrogen in pipelines offers advantages in balancing supply and demand in the energy sector.

If we are to be able to use hydrogen as an energy vector, our present annual production would have to be multiplied many times. Assuming that in the year 2000 some 50% of our energy requirement is for electricity and that 10% of the remainder is for fuel of a synthetic nature, the European Community (of Nine) will have a hydrogen consumption amounting to 30 million tonnes.

---

1. Dr-Ing. G. Beghi, Commission of the European Communities, Ispra Joint Research Centre, Ispra/Varese, Italy

Historical development

The first steps in the new field of hydrogen research were taken at the end of the 1960s, with a view to making use of the high temperature (around 1000 K) of high-temperature gas reactors (HTGR). Studies were undertaken on the design of thermochemical cyclic processes for the decomposition of water into hydrogen and oxygen. Following the energy crisis of 1973, there was a further wave of interest in hydrogen research. In 1974 the IEA[1] was founded. This organization is an autonomous agency within the OECD[2] whose main function consists, in times of crisis, in overseeing contractual commitments, exchanging information and mediating between the oil-producing countries and oil-consuming countries. A subsidiary programme of the IEA is concerned with energy R&D in its member countries.

Numerous research projects have been proposed on a task-sharing basis, and the production and utilization of hydrogen is one of the continuation projects that are being promoted by the IEA. On 6 October 1977 an additional agreement was signed in Paris for a research programme on the development and production of hydrogen from water. The following countries were parties to the agreement: Belgium, Canada, the Commission of the European Communities, Italy, Japan, the Netherlands, Sweden, Switzerland and the USA. The research work was spread over various fields, as follows:

a. thermochemical cyclic processes;

b. water electrolysis (advanced alkaline, solid-polymer electrolyte (SPE) and steam electrolysis);

c. photocatalytic decomposition of water;

d. investigations into hydrogen's future market prospects.

---

1. International Energy Agency

2. Organization for Economic Co-operation and Development

Total expenditure of manpower for the hydrogen programme amounts to some 150 man-years.

Prospects for the future

An initial appraisal of the development of the hydrogen market is based on today's industrial demand for hydrogen as a raw material for fertilizers, plastics and textiles. The "Consult Eur" firm of consultants prepared a forecast of hydrogen consumption for the European Community of Nine in the year 2000, but the results are imprecise because of the uncertain situation of the world markets.

Table 1 shows the development of hydrogen consumption up to the year 2000, on the assumption that consumption will rise by a factor of 2-3.

There is likely to be an increase in hydrogen consumption in hitherto unfamiliar fields, both in the industrial sector and in the energy sector, as in the following examples:

a. as a reducing gas for the manufacture of iron sponge;

b. for the manufacture of synthetic fuels, such as SNG (synthesis gas), methanol and liquid hydrocarbons;

c. direct use of hydrogen as a fuel (process heating gas, and transport fuels for motor vehicles and aircraft).

In 1980 a market study on future production prospects for hydrogen was undertaken on the initiative of the IEA, with eight individual countries taking part (Belgium, Canada, Federal Republic of Germany, Japan, the Netherlands, Sweden, Switzerland and the USA). The development of the hydrogen market in the various sectors (industry, energy) differs in the individual countries because of their differing energy situations.

Table 1 Prospective annual hydrogen consumption in the year 2000 (in 1000 tonnes)

Final consumption	Min	Max	Primary consumption	Min	Max
Fertilizers	2590	3640	Ammonia	2850	4335
Plastics	475	875	Methanol	610	1190
Textiles	145	590	Chemicals	315	840
Oil industry	500	3310	Oil industry	500	3310
Other consumers	565	1260			
Total consumption	4275	9675	Total consumption	4275	9675

It was assumed that between 1985 and 2025 hydrogen consumption will increase by a factor of 12-17, which corresponds to an annual growth rate of 7%.

Although future development prospects are favourable, for the time being the use of hydrogen in these fields is not yet attractive because it is still too expensive. The continuing rise in primary energy prices could alter this situation in the near future. Another important factor is the environmental pollution caused by the use of fossil fuels. Hydrogen would be a true alternative from this point of view.

The use of hydrogen as a fuel gas would be a great improvement environmentally, since the use of fossil fuels increases the carbon dioxide content in the atmosphere. The possible consequences of this concentration of carbon dioxide should not be underestimated.

Despite all the economic uncertainties, even at this stage we are in a position to define possible applications for hydrogen in the future. No time-horizons can be indicated, since differing geographical conditions play a part in this.

The possible areas of use for hydrogen in the future are as follows:

a. production of electrolytic hydrogen, for full-load exploitation of nuclear power stations;

b. direct addition of hydrogen to the existing natural-gas distribution network;

c. use of hydrogen in the processing of heavy oil;

d. use of hydrogen for the manufacture of synthetic liquid or gaseous fuels;

e. direct use of hydrogen as a motor vehicle fuel in urban transport, particularly where air pollution problems are already critical;

f. reduction of iron oxides by means of hydrogen in the steel industry;

g. direct use of hydrogen as an aircraft fuel in air

transport.

These examples give an indication of possible development trends for the future. No firm statements can really be made at this point, since they depend on the particular position in a given case.

Hydrogen technology

The extended use and importance of hydrogen in our energy system depends on the advancement of hydrogen technology. A study of existing and new technologies should as far as possible bring together all the various aspects into a single scheme, along the following lines:

a. Production

In the future we shall not be able to rely on hydrogen processes that start from fossil fuels.

The new processes are based on the following raw materials:

- water and coal;
- electricity or heat.

b. Storage

The storage of hydrogen is of crucial importance for its future uses. The following possible forms of storage are available:

- as a gas: under high pressure in gas cylinders or in underground storage facilities;
- as a liquid: in insulated tanks, cryoadsorbents and metal hydrides.

Transport

At present, most of the hydrogen consumed is used on the spot, where it is produced. In the future, however, pipeline systems for transporting hydrogen will be essential. Such pipelines already exist in various countries (eg Chemische Werke Hüls in Germany has over

200 km of pipeline from Cologne via Düsseldorf, Duisburg and Essen to Dortmund).

Further improvements in transport and distribution systems need to be investigated.

Consumers

Many technical problems have yet to be solved before hydrogen can be used on a large scale as a raw material for industry or as an energy vector.

Safety considerations

Safety considerations play an important part in each of the areas mentioned. These questions need to be studied carefully, since they are just as important for hydrogen as for any other fuel.

All these aspects play an important part in the development of hydrogen as an energy vector. The necessary techniques are already available, and in many instances no new technologies will have to be developed.

The main problems lie in the sphere of economic feasibility, energy policy and the general infrastructure of the energy system.

Summary

This conference is intended to provide a general review of the topics that are associated with hydrogen. It is not possible within so short a time to present a complete and detailed picture of this energy concept. We nevertheless hope that the conference will convey some idea of the following:

a. the present status of hydrogen technology;

b. hydrogen's development potential;

c. the difficulties and the limits of the possible uses of hydrogen;

d. R&D areas.

In conclusion, it should be noted that the development of hydrogen technology is particularly dependent on external factors. The most important factor of all is the future price that will have to be paid for primary energy, and a significant part is played in this by the course of events in the energy policy sector in various parts of the world.

In the long term, hydrogen can make a valuable contribution to an environmentally clean energy system of the future.

The European Community's approach to energy research

by A. Strub[1]

1. Starting-point

Since 1973 the price of crude oil has risen by a factor of 12. Heavy dependence on imported oil makes Europe's economic situation and general development highly vulnerable to factors over which it has no control. What is more, even today our economy is already barely able to afford the cost of the "oil bill" that we have to pay every year.

We have to break free of this excessive dependence on oil.

The countries of the European Community have already taken some initial steps in this direction, but a great deal has yet to be done.

Table 1 shows the development trend of the European Community's energy demand since 1973.

But what the table also shows is that our dependence on oil, which in 1973 was around 59%, has since dropped to 52%. Diversification of our energy supplies and energy conservation measures have borne fruit. In point of fact the Community today imports no more than 475 million tonnes of oil and is able to supply what is now a significant proportion of its needs from its own North Sea oil. Despite this, the bill that the Community has to pay every year for oil imports is continuing to increase, as the following figures demonstrate:

---

1. Dr A. Strub, Commission of the European Communities, Brussels, Belgium

1973	573 million tonnes	10 000 million ECU
1979	475 million tonnes	50 000 million ECU
1980	475 million tonnes	77 000 million ECU[1]

It will be difficult to avoid a further rise in this drain on capital. Certainly, appropriate counter-measures will be more effective if they are taken on the basis of common objectives and a policy harmonized at European level. It is obvious that such a policy can not be defined and adopted by all partners over-night. It takes shape slowly, from the converging or complementary standpoints and programmes of the Member States, whose priorities are naturally - owing to their differing internal situations - not always identical. Nevertheless, it can be observed today that the 10 Member States have already gone most of the way towards agreeing on such a Community energy policy.

2. The Community's energy policy

The Community's energy policy is directed towards the following three imperatives:

1. We must reduce our energy consumption without unduly impairing our economic growth. This demands, above all, strenuous efforts in real energy conservation and the rational utilization of energy.
2. We must open up and develop all potential indigenous energy sources. These include both fossil energy sources (North Sea oil, natural gas and coal) and nuclear fission and the so-called renewable energy sources (solar energy, geothermal energy and nuclear fusion).
3. We must improve, consolidate and, where appropriate, diversify our relations with non-member energy-exporting countries.

Energy R&D can play an important role in satisfying the first two of these imperatives. Hand in hand with economic and fiscal measures (price policy, in-

---

1. ECU ≈ DM 2.5

vestment incentives, etc), R&D will bring about a gradual improvement in the Community's energy situation, at least in the long term.

If a common energy policy is to be of any practical significance, it must not stop short at the formulation of general requirements. It must also include strategic objectives specified in terms of actual figures and time limits. The targets set for the year 1990, on which the competent institutions of the Community have agreed with the Member States, can be summarized as follows:

a. restoring the ratio between primary energy growth and economic growth (GDP) downwards to 0.7 at most;

b. reducing the share represented by oil consumption to some 40% of total energy consumption;

c. increasing the share represented by coal- and nuclear-generated electricity from the present figure of 50% to 70-75% of total electricity generation;

d. intensified promotion of renewable energy sources;

e. reasonable raising or harmonization of energy prices.

These targets have been worked out and agreed on down to the smallest detail and spread between countries. The Member States have undertaken to ensure, through appropriate individual measures tailored to their particular situation, that the collective objectives thus set are achieved. The Commission's task in this is to co-ordinate the national measures and, if necessary, promote them or supplement them with measures of its own.

Table 1 **Energy demand in the European Community**

(in million toe)	1973	1974	1975	1976	1977	1978	1979	1980*
Coal	219	214	192	205	200	202	220	219
Crude oil	555 (59%)	519 (57%)	477 (55%)	508 (55%)	493 (54%)	512 (54%)	525 (53%)	517 (52%)
Natural gas	116	132	140	150	154	161	173	178
Nuclear energy	14	16	19	22	26	28	31	39
Miscellaneous	26	30	33	27	39	35	36	34
TOTAL	930	911	861	911	912	938	985	987

*estimated

3. The Community's approach to energy research

The Commission of the European Communities is conducting a large number of research programmes which, together with parallel efforts at national level, form the technical foundation for achieving the objectives mentioned above.

The Community has been guided by the following four basic principles in choosing its energy research programmes and activities:

1. We must look to a wide variety of technical options and therefore develop them in step with each other. No option should be rejected unless it is quite clearly ineligible on technical, economic or environmental grounds.

2. In contrast to the situation in the past, our future energy supplies can no longer rest on one predominant form of energy. They need to encompass the use of a variety of sources (multiple choice approach), both in order to avoid vulnerable and disproportionate dependence and in order to find, for each region of the Community, the solution best suited to the specific requirement and the local situation.

3. It will take time to make changes to the composition of the energy supply structure. The so-called lead time of new technologies can not be shortened at will. R&D must therefore begin promptly and in a co-ordinated fashion, and this is very much a task for the Community.

4. Because of these lengthy lead times, and also because of the associated imponderables, both technological and political, a long-term energy strategy must allow sufficient margin for setbacks and delays.

A large proportion of energy research is, of course, financed under national programmes or indeed from industry's own resources. This raises the question of the Community's role in energy research. What the

Community is attempting to do is not simply to pursue energy research programmes running parallel to the national programmes, as if it were another Member State. The Community programmes, and the policy actions under which they come into being, are rather aimed at co-ordinating national research activities both vertically and across frontiers, at preventing duplication of effort where it would be fruitless, and at supplementing the national programmes by additional research where the priorities prevailing in an individual Member State would not cover the interests of the Community as a whole. Consequently, before a Community programme in one of the sectors of energy research is set in motion, the Commission checks closely whether the proposed actions do actually satisfy one (or more) of the following conditions or "filters":

a. the action acts as a catalyst (landslide effect) in a field that requires particular impetus (eg certain areas of energy conservation), or

b. the action holds long-term potential for playing a key role in solving our energy problems and, precisely because of its long-term nature or its high cost, patently calls for joint action (eg nuclear fusion, certain areas of solar energy and the utilization of geothermal energy);

c. the project is promising but carries high technical risk that can be kept within bounds by spreading the cost (eg deep-sea oil wells);

d. the project builds on important experience that has been accumulated in a number of regions within the European Community (eg coal research);

e. the research project is to the interest of the entire Community (eg reactor safety, environmental protection, radioactive waste disposal, etc);

f. the project has a pronounced co-ordinating and optimizing effect in regard to national activities (eg plutonium recycling in light-water reactors).

In certain fields energy research has already accomplished results such that the technical feasibility of a process or method is unquestioned, but a further thrust is still required to achieve its commercialization. In cases of this kind the Community also provides financial aid for the execution of what are called demonstration projects, the purpose of which is to demonstrate economic feasibility and provide a stimulus for further applications.

In 1980, the Member States and the Commission itself together spent some 3000 million ECU on energy research, 10% of this expenditure being borne by the Community budget. Table 2 shows the development trend in this expenditure for the years 1976-78. The trend has continued more or less linearly up to the present.

The numerous Community programmes that have been financed and carried out with the above-mentioned 300 million units of account can be divided thematically into the following strategic groups:

a. energy conservation;

b. development of indigenous fossil fuel resources (coal, hydrocarbons);

c. nuclear fission energy;

d. new energy sources (solar energy, nuclear fusion, etc);

e. technologies for the manufacture of synthetic fuels (coal conversion);

f. technologies for the development of new, alternative energy vectors (hydrogen).

The Community is conducting programmes in all the above fields. It can be seen from Table 2, however, that the Community's priorities and expenditure are not proportional to the corresponding national programmes.

From the organizational point of view, the Community uses the following three procedures in carrying out its research programmes:

a. direct action: "in-house research", carried out at one of the Community's own four research establish-

Table 2 Expenditure of the Commission and the Member States on research, development and demonstration in the energy sector*, 1976-1978 (in million u.a)

Subsectors	EUR - 9			COMMISSION			EUR-9 + COMMISSION		
	1976	1977	1978	1976	1977	1978	1976	1977	1978
0. General	6.19	6.76	7.89	-	-	-	6.19	6.76	7.89
1. Energy conservation	101.87	118.24	170.45	0.52	3.10	4.12	102.39	121.34	174.57
2. Petroleum and its derivatives	187.65	281.62	423.07	32.35	47.27	42.69	220.00	328.89	465.76
3. Nuclear energy	998.34	1077.55	1123.62	48.05	61.11	69.58	1046.39	1138.66	1193.20
4. New energy sources and vectors	120.39	153.23	200.78	34.15	52.68	72.08	154.54	203.91	272.86
5. Electricity	126.84	129.91	137.62	-	-	-	126.84	129.91	137.62
6. General studies	27.20	36.46	37.16	1.86	3.72	4.60	29.06	40.19	41.76
T O T A L	1568.48	1803.77	2100.59	116.93	167.88	193.07	1685.41	1971.65	2293.66

*Inventory of 1976-1978 - Energy Research, Development and Demonstration Programmes and Expenditures of the Member States and the European Communities (XII/1000/79-EN)

ments that form the Joint Research Centre (Ispra/Italy, Karlsruhe/Germany, Petten/Netherlands and Geel/Belgium);

b. indirect action: contract research (usually with 50% Community funding), performed by industrial firms, universities and government research bodies in the Member States;

c. concerted action: research which is not financed by the Community but whose content is co-ordinated by the Community, with the Community bearing only the additional costs actually incurred by this co-ordination (cost of meetings, travelling expenses, secretarial services, etc).

Table 3 gives a survey of the Community's current "indirect actions" in the energy sector.

As already explained, it takes some considerable time for a new energy source to be developed to the point where it is ripe for commercialization. Assertions as to the contribution that a given technology will be able to make to energy supplies are subject to very considerable uncertainties. Exhaustive investigations on the subject exist, but will not be discussed in detail here. In a nutshell, however, the indications of these studies are more or less as follows.

In the short term (up to 1985) energy R&D will be able to make only a very limited contribution to Europe's energy supplies. Political and economic policy measures alone can have any tangible effect here.

In the medium term (up to 1995) the die is already cast, technologically speaking: there are in fact only three technological routes by which our energy gap can be closed, namely energy conservation and increased utilization of both coal and nuclear fission energy. Hence the frequently heard reference to the medium-term "COCONUC" strategy. Towards the end of the century, solar energy (solar heating) and geothermal energy will also already be playing a notable role. There is the additional possibility of a contribution from biomass.

Table 3 Community R&D programmes in the energy sector

1.1 Energy			
1.1.1 New energy sources			
- Solar energy	AI	Ⓡ 17,5 M 6 P	46 ME 9 P
- Geothermal energy	AI	Ⓡ 13 M 4 P	18 ME 7 P
- Utilization of hydrogen	AI	Ⓡ 13,24 M 4 P	8 ME 4 P
- Energy conservation	AI	Ⓡ 11,38 M 6 P	27 ME 5 P
- Systems analysis	AI	Ⓡ 3,86 M 7 P	6 ME 9 P
1.1.2 Nuclear fission			
- Plutonium recycling	AI	4,75 ME 3 P	20 ME 9 P (approved by the Commission)
- Radioactive waste management and storage	AI	Ⓡ 19,16 ME 4 P	43 ME 10 P
- Safety in thermal light-water reactors	AI		6,3 ME 3 P
- Decommissioning of nuclear power stations	AI		4,7 ME 3 P
- Reactor development and advanced technology	AI		• 0,66 ME | 0,66 ME
1.1.3 Nuclear fusion			
- Controlled thermonuclear fusion	AI	124 M Ⓡ 113 P	190,5 ME 113 P
- JET project	EC	102,4 ME 180 P	195 ME 150 P
1.1.4 Radiation protection (cf 2.2)	AI	33 M 62 P	59 ME 64 P
1.1.5 Coal			
- Mining techniques	CECA	• 16,5 ME | 17 ME | 17 ME | 18 ME | 18 ME	
- Upgrading of products	CECA	•	

▭ programmes adopted by the Council
▨ programmes approved by the Commission

Ⓡ As at date programme was checked
AI Indirect action
AD Direct action
AC Concerted action
CECA ECSC action
□ Other action
EC Joint undertaking

o Annual budgetary allocations

P Personnel
M million u.a.
ME million ECU

In the long term (after the turn of the century) we can count on three highly promising candidates: fast breeder reactors, solar electricity and thermonuclear fusion.

Thermonuclear fusion - a field to which the European Community is giving very high priority - is not yet advanced to the point where its feasibility can be demonstrated. However, the potential energy contribution of this technology is so enormous that a substantial commitment of resources is justified, even though the risk is correspondingly high. Fast breeder reactors are the logical and necessary consequence of conventional nuclear fission reactors and it is to be hoped that Europe can consolidate its leading role in this field and subsequently bring it to the stage of practical application. Astonishing strides have been made in the last few years in the production of solar electricity using semiconductor cells. With the deployment of research effort on a broad front, it should be possible to introduce the economic large-scale application of this technology towards the turn of the century.

4. An example of the Community's energy research

As an example of the Community's numerous energy research programmes, the indirect action programme on "Energy conservation, new energy sources and energy models" will be described here in some detail.

In 1975, in response to the crisis in energy prices, the Council of Ministers of the European Communities adopted a research programme aimed at employing contract research to promote and accelerate progress in a number of technological fields to which insufficient priority had been given before the crisis. The fields concerned were as follows:

a. energy conservation techniques;

b. the so-called new energy sources (solar energy and utilization of geothermal energy);

c. new energy vectors (hydrogen);

Table 4 The Community's energy research programme (indirect action/contract research)

Topics or subprogrammes	Budgetary resources (million ECU) 1st Programme (1975-79)	2nd Programme (1979-83)
Energy conservation	11.4	27
Hydrogen	13.2	8
Solar energy	17.5	46
Geothermal energy	13.0	18
Energy models and strategy studies	3.9	6
TOTAL	59.0	105

d. systems analysis investigations in the field of energy supplies and strategic studies in the field of energy and energy research.

The first research programme, which covered the period 1975-79, has since been extended by a second programme (1979-1983) with almost double the budgetary allocation. Table 4 indicates the budgetary allocations for the individual subprogrammes. It would take too long here to describe all the research projects (over 700) carried out under the first programme and to present their results in detail. Briefly, however, the work that has been completed can be divided into the following groups of topics:

a. Investigation and Community-wide application of measurement and testing procedures in fields such as investigation of the performance of heat pumps, determination of the efficiency of solar collectors and evaluation of solar heating systems.

b. Investigation of the potential of new energy sources. Examples are: measurement of temperatures throughout Community territory down to depths of 3000 m and their use in the compilation of a "Geothermal Energy Atlas"; measurement of solar radiation in all the Member States (standard calibration of instruments) and publication in an "Atlas of Solar Radiation in Europe"; and determination of the energy potential of the biomass available in Member States (logging wastes, agricultural wastes, etc).

c. Improvement and further development of various energy technologies in the following fields: heat storage, accumulator batteries, water electrolysis plants, solar cells, measuring devices for geothermal drillings, economical vehicle engines, etc.

d. Development of prototypes and pilot plants in the following fields: biomass gasifiers, electrolysis plants, solar-cell power plants, recycling of plastics, etc.

e. Development and construction of large experimental plants; examples are the first European solar power plant (tower principle, electrical capacity 1 MW) and plants for testing solar heating systems (solar pilot test facilities).

f. Execution of case-studies using mathematical models to simulate European energy supplies, and strategic studies on the use of new energy technologies.

Some 100 patents have been applied for on the basis of this work.

The energy research programme described above is an example which illustrates that - in certain fields - Community research leads to more efficient solutions than individual national programmes. Also, Community programmes of this kind axiomatically result in the exchange of results and experience across national frontiers and, in addition, make it possible to utilize results and research capacities that are not always available within one given country. European energy research can in fact make a real contribution to the solution of our energy problems. To sum up, we can say that it is helping to prepare Europe for a situation where not only will oil and gas soon become too expensive for us to afford but there may also no longer be enough of them to go round.

Coal gasification for the production of hydrogen

by K.H. van Heek[1]

1. Introduction

Since the beginning of the industrial era the hydrogen that we need for a variety of purposes, such as chemical synthesis, has been produced by reacting coal or coke with steam (1,2). All the relevant processes consist, in the final analysis, in water molecules splitting into hydrogen and oxygen, with the attachment of carbon to oxygen in the form of CO_2 and the liberation of hydrogen.

In the early days, only coking coals or lignites were used as carbon-containing feedstocks. From the 1930s onwards, however, oil and natural gas also came to be used more and more for the reduction of steam, because they offered advantages over coal in being easier to handle and in many cases cheaper. Today, the corresponding processes such as heavy-oil gasification, naphtha splitting or natural gas predominate in the manufacture of hydrogen. Nevertheless, in a number of countries - and these will be mentioned in more detail later - coal even now still constitutes the basis for the production of hydrogen for liquid-hydrocarbon or ammonia synthesis.

The foreseeable development trend of our future energy situation and the obligation to observe maximum environmental protection in the production, conversion and utilization of energy have given rise to a changed view of ways of producing and utilizing hydrogen in which its more important role in the future will be as an energy vector. When it comes to thinking of the

1. Dr rer.nat. K.H. van Heek, Bergbau-Forschung GmbH, Essen 13

possible feedstocks for hydrogen production in this new scenario, where the quantities needed will be very much larger, the inevitable conclusion is that its production from oil or natural gas can not provide any long-term answer. Of the fossil energy sources, coal alone is regarded as holding great potential here (3). Consequently, alongside other problems associated with safeguarding supplies of energy and raw materials that face the energy expert in the future, the improvement and renewed development of methods for producing hydrogen from coal will play a significant part. A reminder is given below of the various routes to this that are possible, and have even been trodden in the past, in the form of a brief systematic description.

In the production of gas via de-gassing, use is made of the gas that is formed from coal when it is heated to temperatures above 400°C. Coke is left as a by-product and has to be used separately. This process was the basis of the gas works which at one time provided town gas supplies. It is also the basis of the coke-oven technique. The gas produced here has a hydrogen fraction of approximately 50% which can be separated out by adsorptive processes, for example, and used alone.

In the gasification of coal (4-9), there is complete conversion of the organic part of the coal into gas, so that ash alone remains. This is done by reacting the coal with a gasifying agent, eg steam above 700°C.

The following account is concerned solely with coal gasification. The fundamental physical and chemical principles are explained first (10,11), and this is then followed by a discussion of the state of the art and also new developments.

2. Fundemantal physical and chemical principles

The basic reactions are set out in Table 1. The main reaction is the heterogeneous water-gas reaction (1), in which carbon is converted with steam to hydrogen and carbon monoxide. The reaction products can then react either with the gasifying agent or with each

Steam gasification				ΔH kJ/mol
(1)	$C + H_2O$	⇌	$H_2 + CO$	+ 118,5
(2)	$CO + H_2O$	⇌	$H_2 + CO_2$	- 42,3
(3)	$CO + 3H_2$	⇌	$CH_4 + H_2O$	- 206,0
Hydrogasification				
(4)	$C + 2H_2$	→	CH_4	- 87,5
Combustion				
(5)	$C + O_2$	→	CO_2	- 406,0
(6)	$C + \frac{1}{2}O_2$	→	CO	- 123,0

Table 1   Basic reactions of coal gasification

other in two further reactions - the homogeneous water-gas shift reaction (2) and the methanation reaction (3) - with formation of carbon dioxide and methane. Processes (2) and (3) are exothermic, but the endothermic main reaction (1) is dominant, so that overall the gross reaction - consisting of the sum of the three reactions listed under "steam gasification" - is heat-consuming, or endothermic.

Hydrogasification (4), ie the direct reaction of hydrogen with the carbon of the coal, is of great importance for the production of methane as substitute natural gas from coal. Lastly, the reactions with oxygen (5) and (6) also play a part in gasification techniques, on the one hand serving as heat-producing processes in association with endothermic steam gasification but on the other hand also resulting - as in partial combustion (6) - in CO. Reaction (2) is of great importance for the manufacture of pure hydrogen, in the subsequent gas conversion.

The results of some calculations on the thermodynamics of the gasification of carbon with steam (12) are shown in Figure 1. The graphs (temperature dependence for two pressures and pressure dependence for two temperatures) illustrate quantitatively what is to be expected on the basis of the law of least constraint: with rising temperature the proportions of CO_2, CH_4 and H_2O decrease sharply; at 1 bar above 800°C practically only CO and H_2 remain in the reacting mixture. With rising pressure the proportions of non-decomposed water and formed CH_4 increase, while the proportions of hydrogen and CO decrease, which also causes a slight rise in the CO_2. From these relations it can be deduced that for the manufacture of the purest possible hydrogen and CO a relatively low pressure and as high a temperature as possible should be aimed at.

The kinetics of gasification are governed by the interplay of chemical reaction and transport processes (13). Figure 2 groups together some results on the kinetics of the reaction of various fuels with steam (14). It shows the reaction velocity of different coals and cokes with steam at 40 bar as a function of temperature. For all feedstocks, reactivity increases exponentially with temperature. Mainly because of the alkaline earths contained in their ash, which act catalytically, lignites have a very high reactivity. But their chemical and physical structure also plays a part in this. Hard coals and their low-temperature cokes lie in a relatively narrow band well below the reactivities for lignite. It is surprising that differing degrees of coalification, ranging from anthracite to high-volatile bituminous coal, have hardly any effect as regards reactivity. The furnace cokes and pitch cokes that are produced at very high temperatures exhibit extremely low reactivity, which is primarily attributable to their solid structure with the highly ordered arrangement reached at high temperature.

3. The steam gasification process and its potential applications

Figure 3 gives a schematic review of the steam gasification process and its potential applications. Coal and steam are fed into a gas generator, which can be one of three different types (two autothermic and one allothermic) depending on the heat-production technology used. Gas generators classed as autothermic are those in which the heat is produced in the generator itself by an exothermic reaction (generally the combustion reaction). If air/oxygen is used for this,

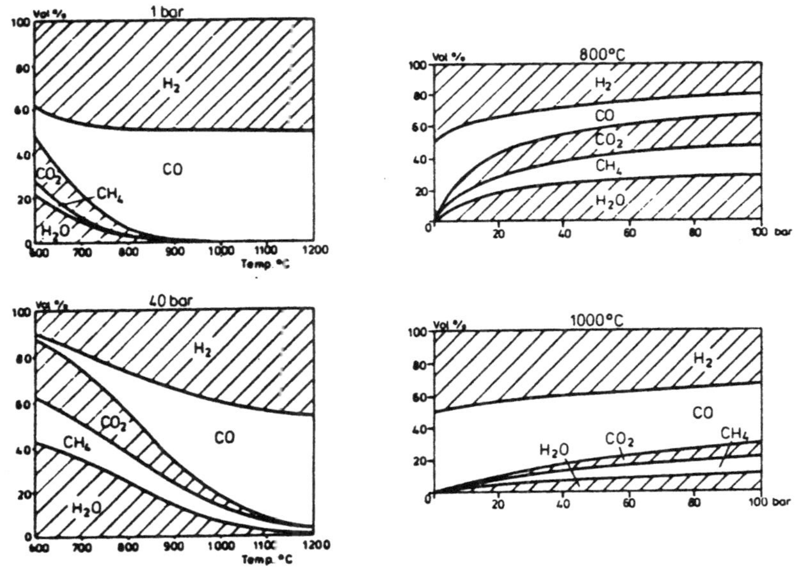

Figure 1 Equilibrium composition of the product gas of the steam gasification of C

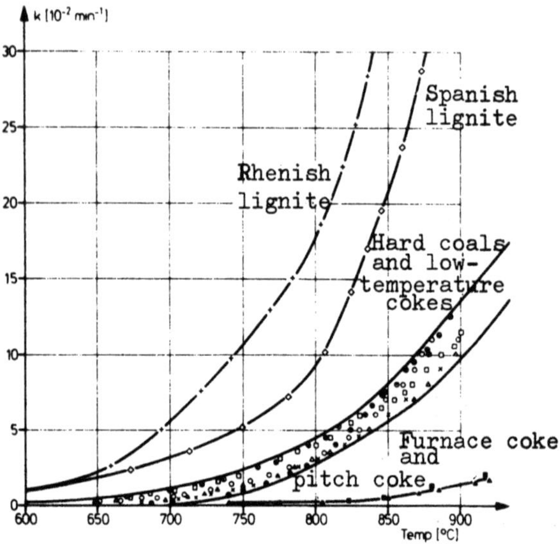

Figure 2 Reactivity of various coals and cokes to steam at 40 bar

the nitrogen introduced along with it causes considerable dilution of the raw gas; the end product is therefore a lean gas with a calorific value of the order of 4000-5000 kJ/m^3 (ntp), which can be used in industrial firing for heating purposes provided that no lengthy transport distances are involved. Examples are smelting or metalworking plants, glassworks, brickworks, lime works and the like, which at one time operated gas generators in Germany itself. This technique is used nowadays in countries where natural gas is not available or can not be distributed because there are no gas networks. For instance, in the German Democratic Republic and in Czechoslovakia there are still, even today, 100 such plants with a total of 400 generators. There are also numerous other plants in operation in developing countries, eg India.

The use of a lean gas in power stations provides the opportunity of first removing, in an intermediate

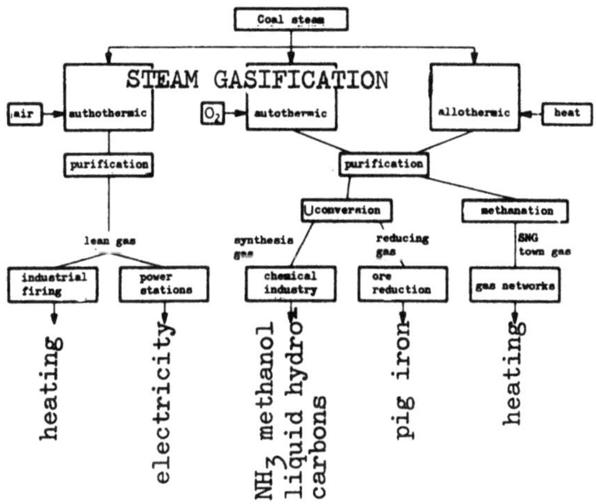

Figure 3 Steam gasification process and its applications

purification stage, all the noxious substances, particularly sulphur, that were originally contained in the coal, and thus producing environmentally clean electricity from coal. In addition, converting coal into gas opens up the possibility of using gas turbines, which greatly improves the efficiency. Work on this technology is therefore in progress in R&D projects in a number of industrialized countries.

Autothermic gasification using pure oxygen gives an undiluted gas consisting essentially of hydrogen, carbon monoxide, methane and carbon dioxide. After purification and conversion, one area of application is to use this to obtain synthesis gas for the manufacture of ammonia, methanol and liquid hydrocarbons. It is mainly for this purpose that coal gasification plants are being operated today, as will be shown later. Another area of use for the gas produced is the direct reduction of ores, particularly iron ores. This technology is primarily of interest for countries that have a poorly cokable coal and can employ the coal gasification route to use such coal for smelting

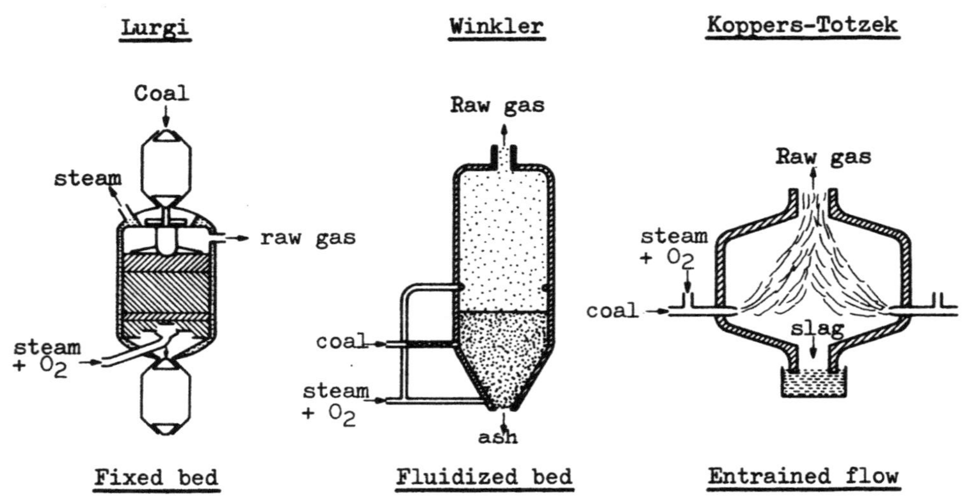

Figure 4 Gas generators for present-day industrial coal gasification

iron ore. In the Federal Republic of Germany, where coal with very good coking properties is available, this route does not really make any sense either technically or economically. Lastly, it is also possible to convert the CO and H_2 produced in an autothermic process into methane in a catalytic reaction in accordance with equation (3), and to distribute this as substitute natural gas or town gas via the gas networks.

Raw gases with lower CO_2 content but of otherwise similar quality are also produced by allothermic gas generators, which differ in that coal, coal products or gasification products are burned outside the generator and the heat is then introduced into the generator using appropriate methods of heat transfer. This obviates the need to produce pure oxygen. However, these processes have not yet won acceptance on a large industrial scale because of their complexity. They will regain importance in the fairly near future by virtue of the fact that it is intended to utilize heat from nuclear high-temperature reactors for gasification, as will be explained in one of the concluding sections of this paper. But this involves breaking completely new ground.

4. Present status of coal gasification

The coal gasification plants used in industry throughout the world today have gas generators of the types illustrated in Figure 4. They are based exclusively on autothermic processes in which coal is gasified with steam and oxygen.

In the Lurgi gas generator (15), gasification is carried out in a fixed bed in which the coal and gasifying agent are introduced in counter-current flow. This results in the formation in the generator of a temperature profile and separate zones, in which with increasing temperature the coal is dried and devolatilized and then gasified and combusted. The process takes place under pressures of between 25 and 35 bar. The coal is therefore fed into the generator via a lockhopper system and the ash is removed via a similar system. The generator is a water-cooled pressurized vessel 3-4 m in diameter.

The fluidized-bed process (16), called the Winkler process after its inventor, was developed for coal gasification on a large industrial scale as long ago as the 1920s. Here, coal is gasified in a fluidized bed with steam and oxygen in a cylindrical gas generator that tapers to a conical shape at the bottom. The coal is fed in near the centre of the bed via a screw conveyor. The gasifying agent is introduced firstly at the bottom and secondly also above the fluidized bed. This ensures that entrained coal particles being carried out with the gas can also be gasified, resulting in a high carbon burn-up and - because of the high temperatures that build up above the fluidized bed - considerable reduction of the by-products. Up till now, the Winkler process has been operated solely at atmospheric pressure.

In the Koppers-Totzek entrained-flow reactor (17), pulverized coal is gasified with oxygen and steam in an entrained flow of coal dust. Because of the short residence times, the introduction of relatively large quantities of oxygen produces very high temperatures and high conversion rates are therefore achieved. The ash runs down the walls as liquid slag and is removed from beneath the generator. Directly above the generator there is a waste-heat boiler for heat recovery.

Table 2 makes it possible to compare these three currently used coal gasification processes. The most important technical data on the generators have already been given in the text. As regards the feed coal for the process, it should be noted that the Lurgi and Winkler processes are suitable at best for moderately caking coals, whereas caking properties are immaterial with the Koppers-Totzek process. Also, the fixed-bed process does not tolerate fine-grained material: only lump coal can be used, and fines have to be diverted to some other use suoh as a power station for the generation of electricity and steam for coal gasification. However, it is only fair to record that future developments are aimed at broadening

Gas generator		**Lurgi**	**Winkler**	**Koppers-Totzek**
	type	fixed-bed	fluidized-bed	entrained-flow
	temp. (^oC)	700-1000	approx. 1000	>1300
	pressure (bar)	35	1	1
	gas production (m^3/h)	35-50 000	17-20 000	20 000 (2 nozzles) 50 000 (4 nozzles)
Coal	rank	lignite and hard coal up to moderately caking	lignite and hard coal up to moderately caking	all coals
	grain size (mm)	6-40	0-8	<0.1
Raw gas	vol.% H_2	37-39	35-46	31
	CO	20-23	30-40	58
	CO_2	27-30	13-25	10
	CH_4	10-12	1-2	0.1
Other criteria		- good heat utilization - little discharge - high carbon content - tar and oil - high steam consumption	- no tar and oil - high carbon content - high discharge - low degree of gasification	- no by-products - high steam decomposition - low carbon content - high coal and oxygen consumption - high discharge

Table 2 Comparative data on coal gasification processes in industrial use

the feed coal tolerance. As regards the gas composition, it should be pointed out that in the fixed-bed process there is a markedly higher methane content in the raw gas, which is attributable both to the separate zones that form inside the gasifier and to the elevated pressure. Consequently, the Lurgi process is particularly well suited to the production of a high-heating-value gas and was in fact originally developed for the production of town gas. The raw gas from the Winkler and Koppers-Totzek processes, in contrast, exhibits a very low methane content and is therefore very suitable for conversion into pure hydrogen. The table also lists a number of criteria relevant to the choice of process according to the rank and cost of the coal and the target product.

A major field of application of coal gasification is in the manufacture of ammonia for fertilizer production (18). At present, some 3.8 million tce are used annually for this purpose in various developing countries. Most such plants are based on hard coal, and the process most widely used is the Koppers-Totzek process. The Lurgi process is used in South Africa for the gasification of hard coal with a wide range of products (19). A plant with 17 Lurgi gasifiers has been in operation there since the early 1960s: 2.5 million tonnes of hard coal are gasified annually, from which 270 000 tonnes of liquid hydrocarbons are then obtained by Fischer-Tropsch synthesis. Chemical feedstocks and town gas are produced at the same time. A second plant with twice the capacity came into operation in 1980, and a third plant of the same size is under construction.

5. Further development of gasification processes

Intensive work is in progress in the Federal Republic of Germany, the USA and other countries on the further development of coal gasification for future applications (20, 21); the lines of investigation being pursued can be summarized as follows:

a. continued development of conventional processes with a view to broadening feed coal tolerance, increasing efficiency and, for entrained-flow and fluidized-bed gasification, using higher operating pressures;

b. reduction of coal consumption by using nuclear process heat from high-temperature reactors;

c. production of methane as substitute natural gas via hydrogasification;

d. use of gasification for environmentally clean generation of electricity.

From the standpoint of hydrogen production, only the first two of these objectives enter into our discussion. The German development projects for the further development of conventional processes and for the use of nuclear process heat from high-temperature reactors are specifically relevant here, and will therefore feature prominently in the following description.

Table 3 lists the activities in progress in the Federal Republic of Germany on the further development of conventional gasification technology. The first column shows the abbreviated name of the project together with the firm responsible for the development. It can be seen that the coal producers have an important stake in the development of gasification processes. With the exception of the High-Temperature Winkler (HTW) process and hydrogasification, all the processes being developed are for the gasification of hard coal. As regards the type of gas generator, it can be noted that there are two fixed-bed processes, three fluidized-bed processes and three entrainment processes in course of development. All the experimental plants are in operation. With the exception of the Shell-Koppers process, all the projects are backed by public funding. Some of them will now be described in detail.

Table 3 Gasification projects in the Federal Republic of Germany

Name of project	Operator	Technical description	Status and objectives
Fixed-bed pressure gasification (Lurgi)	Ruhrgas Ruhrkohle Steag	Fixed-bed gasification of hard coal with steam and oxygen, separate removal and processing of devolatilization and gasification gas 100 bar, 700-1000°C, 7 t/h	Experimental plant in operation since 1979 at Dorsten. Further development of the Lurgi process to high pressures for SNG production
Pressure gasification of coal dust (Texaco)	Ruhrkohle Ruhrchemie	Entrained-flow gasification of hard coal in a coal/water suspension with oxygen 40 bar, 1300-1500°C, 6 t/h	Experimental plant in operation since 1978 at Oberhausen Further development of the Texaco process for the manufacture of synthesis gas, plus utilization of hydrogenation residues
Saarberg-Otto process	Saarbergwerke Dr C. Otto	Entrained-flow gasification of hard coal in a slag-bed generator with O_2 30 bar, 1500-1700°C, 10 t/h	Experimental plant in operation since 1979 at Völklingen, Saarland Further development of the Rummel generator for pressurized operation in the production of synthesis gas and fuel gas with later operation with air
Shell-Koppers process	Shell Krupp-Koppers	Entrained-flow gasification of hard coal with oxygen and steam 30 bar, 1500-1850°C, 6 t/h	Experimental plant in operation since 1979 at Hamburg-Harburg Further development of the Koppers-Totzek gas generator for pressurized operation in the production of synthesis gas and fuel gas with later operation with air
Gasification of high-inert coal	Kohlegas-Nordrhein	Fixed-bed gasification of shaped high-ash hard coals in intermittent operation (blow-or-make): hot blowing with air, gasification with steam 1-6 bar, 950-1150°C, 1.6 t/h	Experimental plant in operation since 1979 at Hückelhoven Production of fuel and synthesis gas from coal with a high inert content
High-Temperature Winkler (HTW)	Rheinische Braunkohlenwerke	Fluidized-bed gasification of lignite with oxygen and steam, with addition of Ca to raise the ash melting point and as S-acceptor 10 bar, 1100°C, 1 t/h	Experimental plant in operation since 1979 at Frechen-Wachtberg Production of synthesis and reducing gas from lignite
PNP[1] hydrogasification	Rheinische Braunkohlenwerke and PNP partners	Fluidized-bed gasification of coal with H_2; 0.1 t/h (H_2 production by CH_4 reforming in a tube furnace heated with nuclear process heat)	Experimental plant in operation since 1975 at Wisseling. Production of SNG using process heat from nuclear high-temperature reactors 10 t/h (dry basis) under construction
PNP steam gasification	Bergbau-forschung and PNP partners	Fluidized-bed gasification of coal with steam with coupling of nuclear process heat 40 bar, 800°C, 0.2 t/h	Experimental plant in operation since 1976 at Essen Production of synthesis gas and SNG using process heat from nuclear high-temperature reactors

1. Prototype Nuclear Process Heat Plant
PNP partners: Bergbau-Forschung, Essen; Kernforschungsanlage, Jülich; Gesellschaft für Hochtemperaturreaktor-Technik mbH, Bensberg; Hochtemperatur-Reaktorbau, Mannheim; Rheinische Braunkohlenwerke, Cologne

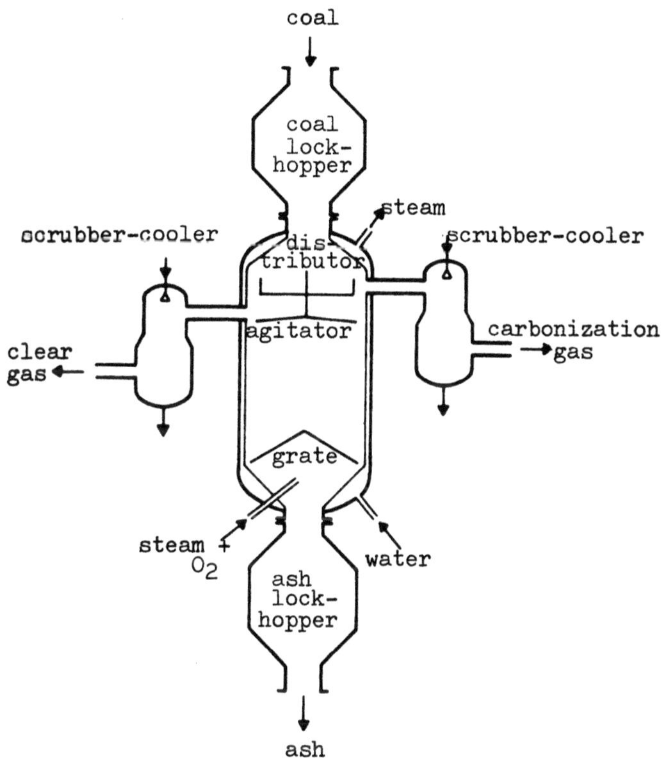

Figure 5 Generator for fixed-bed pressure gasification at pressures of up to 100 bar: Lurgi "Ruhr 100" type

Figure 5 shows the gas generator for fixed-bed pressure gasification up to pressures of 100 bar. The generator essentially follows the operating principle of the conventional Lurgi generator; the most striking modification, however, is that there are two separate gas outlets, one above the gasification zone from which a clear gas can be taken off and the other above the devolatilization zone from which a carbonization gas can be taken off. The latter contains the impurities from the pyrolysis process, ie essentially the tars, which can be more easily separated from the smaller quantity of carbonization gas than from the total gas in the case of the conventional Lurgi process, and then processed. A special agitator design is incorporated to enable even strongly caking coals to be processed in the generator. The experimental plant at Dorsten was brought into operation in the autumn of 1979. Two other further developments of the Lurgi gasifier which do not form part of the German gasification projects should also be mentioned (15).

In the last few years, at the former pressure gasification plant at Westfield in Scotland, a fixed-bed generator incorporating the slagging principle has been under trial with backing from private US interests. By using an increased input of oxygen, such a high temperature is reached in the lower section of the generator that the ash can be tapped off in the form of liquid slag. The hope is that this will bring a significant increase in the conversion rate. A generator of this type is planned for a demonstration plant in the USA. The second of the abovementioned developments is an increase of the generator diameter to 5 m, corresponding to a gas production of 100 000 m^3/h. A first prototype is under trial at Sasol in South Africa.

Figure 6 shows a flow diagram of the Texaco gasification process (23).

Coal is finely ground and slurried with water. This suspension is pumped into the gas generator, which is under pressure, and the coal is gasified there in an entrained flow with oxygen and the steam produced from the slurry. The reaction takes place at temperatures above 1300°C. This causes the formation of liquid slag, which drops into a water bath and is then removed. The waste heat from the high-temperature raw gas is utilized via steam recovery, and the gas is purified and passed on to its intended use. The experimental plant on the Ruhrchemie premises at Oberhausen-Holten has been in operation since the end of 1977.

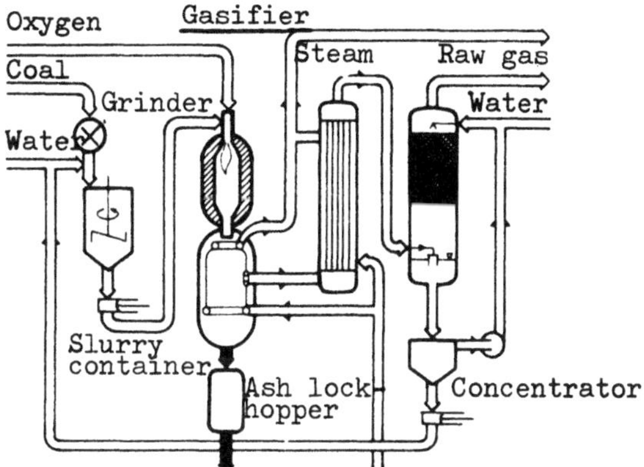

Figure 6 Coal-dust gasification under pressure by the Texaco process

The Saarberg-Otto process (24) also is a coal-dust gasification process, characterized in this case by the fact that the reaction takes place in a slag-bath generator, a principle that has already been tested on a fairly large scale in an earlier Rummel generator at atmospheric pressure. As Figure 7 shows, in this case the finely ground coal, in a gaseous suspension with oxygen and steam, is blown into the lower part of a cylindrical gas generator, where the slag bath is situated, and gasified at very high temperature. At the exit from the generator, cold raw gas is injected to cause the entrained drops of slag to solidify. These are then separated out in a cyclone and, as long as they still contain carbon, are recycled to the gasification zone. As in the Texaco process, intensive cooling of the raw gas then takes place with recovery of its sensible heat in the form of steam. The experimental plant for this process is operated at Saarberg's own Fürstenhausen coking works near Völklingen.

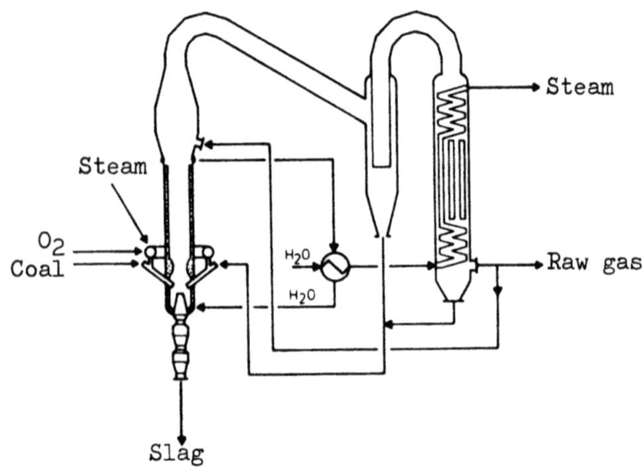

Figure 7 Coal gasification under pressure by the Saarberg-Otto process

6. Coal gasification using nuclear process heat

In all conventional processes the coal fulfils a dual function: it not only supplies the raw material for the gas that is to be produced but is also the energy source for the heat of reaction, the heat for heating up coal and steam and all other forms of energy needed within the plant. In practice, this means that 30-40% of the coal has to be burned in order to convert the remainder into gas. By using process heat from nuclear high-temperature reactors (HTR) (26) the energy requirement of a gasification plant can be covered: all the coal is therefore converted into usable gas. The advantages of this procedure are conservation of coal reserves and more efficient utilization of the available coal (whether this is from indigenous resources or imported), reduction of coal-specific atmospheric pollution (particularly CO_2), since less coal is burned, more efficient utilization of nuclear energy by coupling the heat into a chemical process, and lower production costs for coal gas.

To bring such processes to the point where they are ready for large-scale industrial use, work is being carried out jointly within the PNP project (Prototype Nuclear Process Heat Plant) by a consortium of firms in the nuclear industry and the coal industry* to develop the HTR, heat-exchanging components and appropriate gasification processes - hydrogasification and steam gasification (27).

For the production of CH_4 the route being pursued is the combination of hydrogasification with nuclear-heated methane reforming, and for the production of synthesis gas and hydrogen the route is steam gasification (28). The latter is described in some detail below.

1 Nuclear high-temperature reactor
2 Heat exchanger
3 Steam generator
4 Gas generator
5 Gas processing

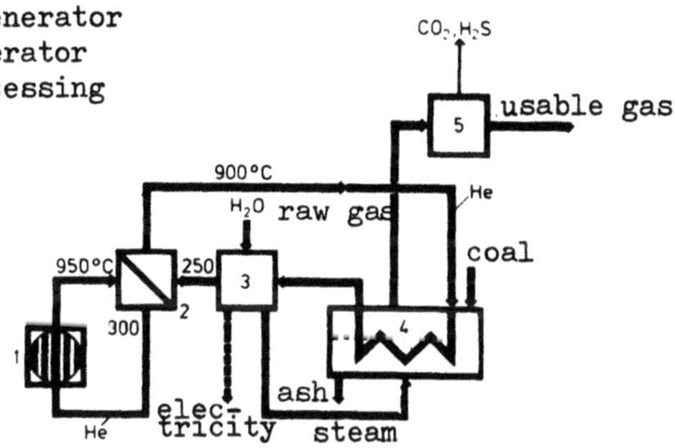

Figure 8 Coupling of a nuclear high-temperature reactor and steam gasification

---

*Bergbau-Forschung GmbH, Essen; Gesellschaft für Hochtemperaturreaktor-Technik mbH, Bensberg; Hochtemperatur-Reaktorbau GmbH, Mannheim; Kernforschungsanlage Jülich GmbH, Jülich; Rheinische Braunkohlenwerke AG, Cologne

Figure 8 shows the coupling of a nuclear high-temperature reactor and steam gasification. The reactor is the pebble-bed type developed in Germany, with spherical fuel elements in a moving bed through which helium is passed, the helium being thereby heated to a temperature of up to 950°C. An experimental reactor has been in operation for years at the Kernforschungsanlage Jülich. For safety reasons, the heat is coupled via a heat exchanger to a secondary loop, in which helium is pumped round. This passes through a heat exchanger in the gas generator whose heat-transfer tubes are immersed in a fluidized bed of coal and steam and thus supply the heat for the endothermic reaction. At a lower temperature level, the heat of the helium is used to generate steam, part of which is employed as process steam and part for electricity generation. The raw gas produced is converted and used in the conventional manner. The gas generator required is a horizontal cylindrical pressurized vessel 33 m long and 7 m in diameter. In the lower part there is a trough into which steam can flow through openings from beneath into the coal/steam fluidized bed supported by it. Coal or char is fed into the fluidized bed at one end and then moves through the fluidized current with an attainable carbon burn-up of over 95%, so that at the other end a high-ash fraction can be removed.

Fluidized-bed gasification with coupling of heat from a helium loop at pressures of up to 40 bar and helium entry temperatures of between 900 and 1000°C is being carried out for the first time in a Bergbau-Forschung semi-industrial plant (21). Figure 9 shows the flow diagram: the gas generator is a section of the industrial-scale gas generator, so that the height of the fluidized bed and the length of the heat-transfer tubes in fact correspond to the industrial-scale design. The fluidized bed itself has an area of approximately 1 m^2 and can operate up to a height of 4 m. The plant has been in operation since mid-1976.

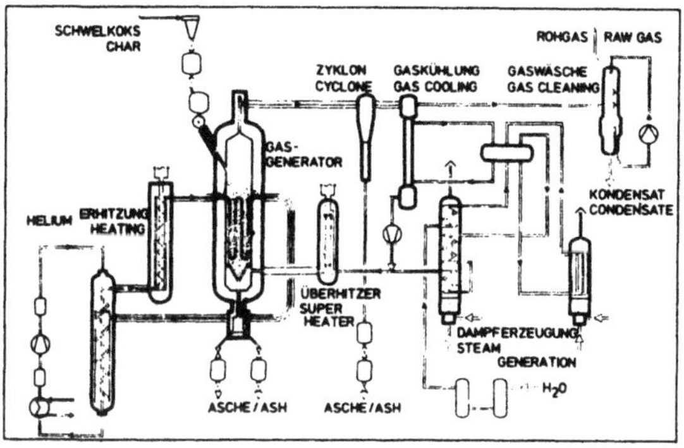

Figure 9 Flow diagram of semi-industrial plant for the steam gasification of coal with coupling of heat from an electrically heated helium loop

7. Prospects for hydrogen production by coal gasification

For future application, the first question that arises is: which process is best suited to the production of synthesis gas or hydrogen? As Table 4 shows (18), fluidized-bed and entrained-flow reactors offer special advantages for this, in that as a result of the high temperatures the raw gas, freed from CO_2, consists almost solely of H_2 and CO, and the latter can be completely converted to H_2 in a subsequent conversion stage.

The next question is: when will the coal gasification processes be fully developed ready for use? The conventional processes are already available: calculating on the basis of a construction time of five years, industrial-scale plants (ie with an annual coal throughput of several million tonnes) could in theory be operational from 1986 onwards. In contrast to this, all the pilot plants that we have discussed presumably still have to pass through several years of

Table 4 Potential of selected gasification processes for the production of specific product gases (30)

PROCESS	MAIN PRODUCT Fuel gas 4	Synthesis gas 11.5	Town gas 18	SNG 38 MJ/m³	MINIMIZATION OF BY-PRODUCTS
Fixed-bed					
Lurgi: conventional	+	+	++	+	-
slagging	±	+	++	+	+
Ruhr 100	-	-	++	++	+
Gasification of high-inert coal	-	+	+	±	++
Fluidized-bed					
Winkler: conventional	++	++	±	±	+
High-Temperature Winkler	±	++	±	±	+
Nuclear coal gasification:					
steam gasification	-	++	+	+	+
hydrogasification	-	±	±	++	+
Entrained-flow					
Koppers-Totzek	-	++	±	±	++
Shell-Koppers	+	++	±	±	++
Texaco	-	++	±	±	++
Saarberg-Otto	+	++	±	±	++

- not possible; ± possible; + suitable; ++ particularly developed for such

experimental operation and the demonstration-plant stage before they can be brought into use for commercial gas production. The late 1980s would therefore be the earliest date at which they could start to be a serious consideration. And as regards nuclear gasification, it will not be until towards the end of the 1990s that its technical development will be sufficiently advanced for the operation of large-scale plants in which it can make a notable contribution to our supplies of energy and raw materials.

The question of the cost of producing hydrogen by gasification of indigenous hard coal can be answered by examining Table 5 (31). The costs given in the table for synthesis gas are essentially the same for hydrogen. It shows that, at present, the utilization of German hard coal is not yet competitive with other feedstocks for hydrogen production. However, the heavy-oil utilization costs are already close to those for coal. Higher investment costs alone, necessitated by the technical difficulties of handling coal, account for the difference in the hydrogen production costs. But with the advent of the expected shortage of petroleum and natural gas the balance will shift more in favour of coal. It should be mentioned that German lignite is already fully competitive today as a feedstock for hydrogen production. A fairly large-scale plant for the production of synthesis gas is under construction in the Rhine district.

For a successful application of coal gasification, it is vital that we should have enough hard coal available for the purpose into the fairly distant future. The German coal industry believes that in the 1990s it will be able to supply 30-40 million tonnes/year for gasification and liquefaction. Comparison with the primary energy balance sheet, which includes an annual 150 million tce in the form of petroleum, shows a possible shortfall that will not be covered by oil and gas from coal. Imported coal represents another possibility. Priority must therefore be given to the development of those processes which promise optimum coal utilization.

Table 5 Comparison of synthesis gas costs (31) (DM/1000 Nm^3) (1979 prices)

Feedstock	Heavy oil	Natural gas	Hard coal
Specific investment costs	252	166	324
Fuel costs	125	102	129
Capital costs	50	33	80
Operating costs	7	9	9
By-product credits			
Production costs	182	144	218

From this point of view, coal gasification using nuclear process heat from high-temperature reactors offers particularly significant potential for the necessary improved efficiency in the utilization of our reserves of available primary energy and can therefore play a decisive role in safeguarding energy supplies.

References

(1) P. Dolch: Wassergas (Water gas), Leipzig 1936, J.A. Barth Verlag

(2) H.W. von Gratkowski: Kohlevergasungs-Verfahren (Coal gasification processes), in Ullmanns Encyklopädie der technischen Chemie 10 (1958), pp 376 ff, and supplementary volume (1970), p 389

(3) W. Peters: Die zukünftige Bedeutung der Kohle als Rohstoff und Energieträger (The future importance of coal as a raw material and energy vector), Glückauf 114 (1978), pp 330-339

(4) J. Meunier: Vergasung fester Brennstoffe und oxidative Umwandlung von Kohlenwasserstoffen (Gasification of solid fuels and conversion of hydrocarbons with oxygen), Verlag Chemie GmbH, Weinheim /Bergstrasse, 1962

(5) J. Schmidt: Technologie der Gaserzeugung, Band II, Vergasung (Technology of gas production, Vol II,

Gasification), Leipzig 1966, VEB Deutscher Verlag für Grundstoffindustrie
(6) H.-D. Schilling, B. Bonn and U. Krauss: Kohlevergasung (Coal gasification), Rohstoffwirtschaft International Vol 4, with an introduction by W. Peters, Verlag Glückauf, Essen, 1979
(7) J. Falbe (Ed): Chemierohstoffe aus Kohle, Kap.5, Kohlevergasung (Chemical feedstocks from coal, Chapter 5, Coal gasification), Georg Thieme Verlag, Stuttgart, 1977
(8) H. Jüntgen and K.H. van Heek: Grundlagen, Anwendung und Weiterentwicklung der Kohlevergasung (Principles, application and further development of coal gasification), Gas- und Wasserfach - Gas/Erdgas, Vol 120 (1979), p 559 and Vol 121 (1980), p 6
(9) H. Jüntgen and K.H. van Heek: Kohlevergasung - Grundlagen und Technische Anwendung (Coal gasification - principles and industrial application), Verlag Karl Thiemig, Münich, to appear in autumn 1981
(10) K. Hedden: Kohlevergasung, physikalisch-chemische Grundlagen (Coal gasification - physical and chemical principles), in Ullmanns Encyklopädie der technischen Chemie, Vol 10 (1958), pp 362 ff
(11) H. Jüntgen: Theoretische Grundlagen von Vergasungsprozessen (Theoretical principles of gasification processes), in Kohlevergasung in der Energietechnik, VGB-Konferenz, VGB-TB 201, VGB-Kraftwerkstechnik GmbH, Essen 1979
(12) I. Barin and D. Neuschütz: Berechnung der Gleichgewichtszusammensetzungen und Russgrenzen im System C-CH_4-H_2-CO-H_2O-CO_2 (Calculation of the equilibrium compositions and soot limits in the C-CH_4-H_2-CO-H_2O-CO_2 system), Archiv für Eisenhüttenwesen 46 (1975), p 159
(13) E. Wicke, K. Hedden and M. Rossberg: Beiträge der reaktionskinetische Forschung und Technik der Vergasung und Verbrennung (Reports on the reaction-kinetics research and technology of gasification and combustion), Brennstoff-Wärme-Kraft 8 (1956), p 264
(14) K.H. van Heek, H. Jüntgen and W. Peters: Fundamental studies on coal gasification in the utili-

zation of thermal energy from nuclear high-temperature reactors, Journal of the Institute of Fuel 249 (1973), pp 249-258

(15) H. Bierbach and H. Jockel: Weiterentwicklung der Lurgi-Druckvergasung (Further development of Lurgi pressure gasification), Stahl und Eisen 100 (1980), No 7, pp 371-376

(16) J. Anwer and F. Bögner: Kohlevergasung im Fluidatbett unter Druck (Coal gasification in a fluidized bed under pressure), Brennstoff-Wärme-Kraft 28 (1976), pp 57-60

(17) H. Staege: Die grosstechnische Anwendung der Kohlevergasung nach Koppers-Totzek (Large-scale industrial application of Koppers-Totzek coal gasification), Erdöl-Erdgas Z. 95 (1975), No 11, pp 390-397

(18) K. Chari: Coal-based ammonia plants, Chem. Age India 28 (1977), p 698

(19) J.C. Hoogendorn: Die neue Anlage zur Erzeugung flüssiger Kohlenwasserstoffe aus Kohle in Südafrika (The new plant for the production of liquid hydrocarbons from coal in South Africa), Gas, Wärme Internat. 25 (1976), p 283

(20) K.H. van Heek: Überblick über den internationalen Entwicklungsstand der Kohlevergasungsverfahren (Review of the international stage of development of coal gasification processes), Stahl und Eisen 100 (1980), p 363

(21) Ch. Brecht, H.W. von Gratkowski and G. Hoffmann: Vergasung und Hydrierung von Kohle (Gasification and hydrogenation of coal - A tabulated review of domestic and foreign developments and processes in large-scale industrial use), Gas, Wärme International 29 (1980), pp 367-387

(22) C. Lohmann and G. Röbke: Das Kohlevergasungsprojekt Dorsten Lurgi - Ruhr 100 - Bisherige Betriebserfahrungen (The Lurgi Ruhr 100 coal gasification project at Dorsten - Operating experience to date), Gas- und Wasserfach/Gas 121 (1980) p 359

(23) B. Cornils, J. Hibbel, J. Langhoff and P. Ruprecht: Stand der Texaco-Kohlevergasung in der Ruhrchemie /Ruhrkohle-Variante (Status of the Ruhrchemie/ Ruhrkohle version of Texaco coal gasification),

Chem.-Ing.-Tech. 52 (1980), p 12

(24) M. Rossbach, A. Meyer and V. Hörnung: Das Saarberg/Otto-Kohlevergasungsverfahren (The Saarberg-Otto coal gasification process), Stahl und Eisen 100 (1980), p 383

(25) A. Linke and E.V. Vogt: Die Druckvergasung von Kohle im Flugstrom nach Shell-Koppers (Entrained-flow gasification of coal under pressure by the Shell-Koppers process), Chem.-Ing.-Technik 52 (1980), No 9, pp 742-745

(26) L. Hardt: Stand und Potential der Entwicklung von Hochtemperaturreaktoren (Status and potential of the development of high-temperature reactors), Glückauf 113 (1977), No 11, pp 545-551

(27) E. Arndt, R. Fischer, W. Fröhlich, H. Jüntgen, H. Teggers and I. Weisbrodt: Synthetisches Erdgas aus Kohle und Hochtemperatur-Wärme (Synthetic natural gas from coal and high-temperature heat), Erdöl und Kohle, Erdgas, Petrochemie 32 (1979), p 17

(28) K.H. van Heek: Wasserdampfvergasung von Kohle unter Einkopplung nuklearer Prozesswärme aus Hochtemperatur-Kernreaktoren (Steam gasification of coal with coupling of nuclear process heat from high-temperature reactors), in Kohlevergasung in der Energietechnik, VGB-Konferenz, VGB-TB 201, VGB-Kraftwerkstechnik GmbH, Essen 1979

(29) R. Kirchhoff, K.H. van Heek and H. Jüntgen: Betrieb einer halbtechnischen Versuchsanlage zur allothermen Druckvergasung von Kohle und Wasserdampf (Operation of a semi-industrial experimental plant for the allothermic pressure gasification of coal and steam), Compendium 80/81 Erdöl und Kohle, Erdgas, Petrochemie, supplementary volume, pp 177-179

(30) H. Jüntgen and K.H. van Heek: Bewertungskriterien für die Vergasungsverfahren und Einsatzkohlen (Evaluation criteria for gasification processes and feed coals. 1. Factors in the selection of gasification processes according to the coals available and the target products), Compendium 80/81 Erdöl und Kohle, Erdgas, Petrochemie, supplementary volume, pp 171-173

(31) H.-D. Schilling, D. Wiegand and B. Strobel: Wirtschaftliche Perspektiven der Steinkohleverflüssigung in der Bundesrepublik Deutschland (Economic prospects of coal liquefaction in the Federal Republic of Germany), Energiewirtschaft, 1979, No 4, pp 260-267

The electrolytic production of hydrogen

by H. Wendt[1]

Introduction

The electrolysis of water is of interest as an economic industrial-scale method of producing hydrogen in cases where (i) fossil fuels such as petroleum, natural gas and coal (from which hydrogen can also be obtained) are scarce or expensive, or both, and (ii) electrical energy from sources other than fossil fuels is available in sufficient quantity and relatively cheaply.

This situation will arise in the foreseeable future in the energy economy of France, and probably also that of Belgium, inasmuch as nuclear power stations will represent more than 50% of the total power station capacity.

Nuclear power stations have to be operated with a load which is constant in time; hence, during the seasons or times of day when there is a decreased demand for electricity the energy produced is particularly cheap and could be used to produce hydrogen.

So far, water electrolysis technology has been developed to the point where extremely robust and reliable electrolysers are available which operate without excess pressure or under a pressure of 30 bar. However, most electrolysers operate with relatively high cell voltages (between 1.8 and 2.2 V), ie with only up

---

1. Prof. Dr H. Wendt, Institut für Chemische Technologie, Technische Hochschule Darmstadt, and project leader of the European Community's "Water Electrolysis" R&D programme

to % electrical energy utilization; furthermore, these electrolysers are operated with relatively modest current densities of 0.15-0.30 A/cm^2.

High cell voltages imply high energy costs, and low current densities imply that the electrolysers give small yields per unit time, leading to relatively high capital costs. For these reasons hydrogen produced electrolytically is at present twice as expensive as hydrogen produced from fossil fuels.

There has not been any real economic incentive for continuing to improve the state of the art, since up till now no sufficiently large market for water electrolysers has existed.

Figure 1 compares the thermodynamic data (free enthalpies converted into hypothetical cell voltages) with the present state of the art and also shows the projected target envisaged for a further development of water electrolysis. Figure 2 again shows the present state of the art and the target for development, this time in economic terms.

Figure 3 illustrates the three main reasons for the relatively unsatisfactory performance of present-day water electrolysers:

a) powerful inhibition of the electrode reactions by the oxygen evolved at the anode and the hydrogen evolved at the cathode, leading to anodic and cathodic overvoltages;

b) too large a potential drop between the electrodes, caused by too wide a distance between the electrodes and too high a resistance of the diaphragm per unit area;

c) an increase in the internal resistance of the electrolytic cell caused by the gases produced at the anode and cathode.

From this, the following focal points can be singled out for R&D work designed to produce new water electro-

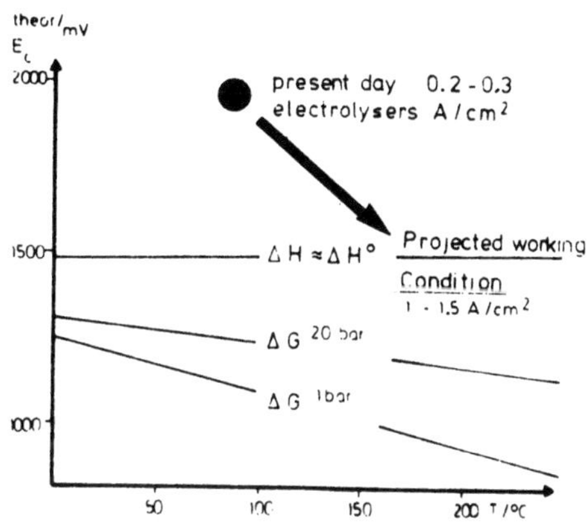

Figure 1 Hypothetical cell voltages calculated from the enthalpy (ΔH) or free enthalpy (ΔG) for the decomposition of water. Present state of the art and target for development in advanced electrolysis

lysers with higher performance that will reduce the expenditure on energy and capital costs for water electrolysis:

a) discovery of new electrocatalysts for the evolution of hydrogen and oxygen, and attempts to promote the electrode reactions thermally by raising the working temperature;

b) development of new diaphragms having a mechanical stability at least as good as asbestos diaphragms, but a lower resistance per unit area;

c) modifications of the electrode geometry and the internal configuration of the electrolyser so as to achieve an appreciable decrease in the unwanted ohmic loss and to eliminate the effect of bubbles.

Present state of the art	Development target
Production cost of hydrogen	
DM 0.42/Nm^3 H_2	DM 0.30/Nm^3 H_2
Energy consumption	
4.5 kWh/Nm^3 H_2	3.8 kWh/Nm^3 H_2
Capital costs	
DM 0.10-0.14/Nm^3 H_2	DM 0.08-0.12/Nm^3 H_2
Rate of production per m^2 of electrode surface	
1 $Nm^3/m^2 \cdot h$	4.5 $Nm^3/m^2 \cdot h$

Present state of the art - Development target

Figure 2

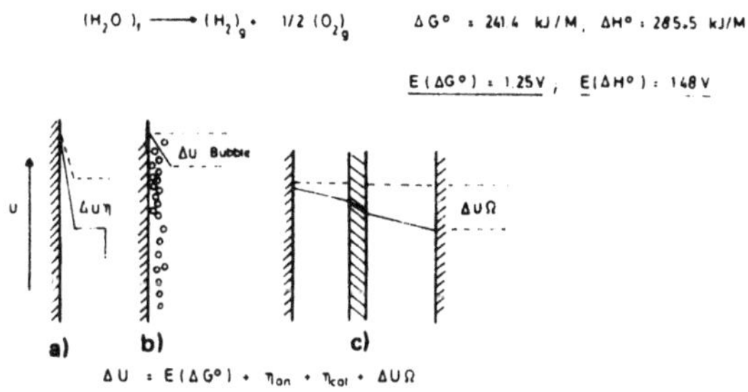

present state: 30 wt.% KOH, $80^{\circ}C$, 0.2-0.3 A/cm^2

$\Delta U \approx 1.8\text{-}2.1$ V

target: 1 A/cm^2

$\Delta U \approx 1.4\text{-}1.5$ V

<u>Figure 3</u> Causes of excessive cell voltages:
a) cathodic and anodic overvoltages,
b) masking effect of bubbles,
c) ohmic potential losses within the cell and in the diaphragm

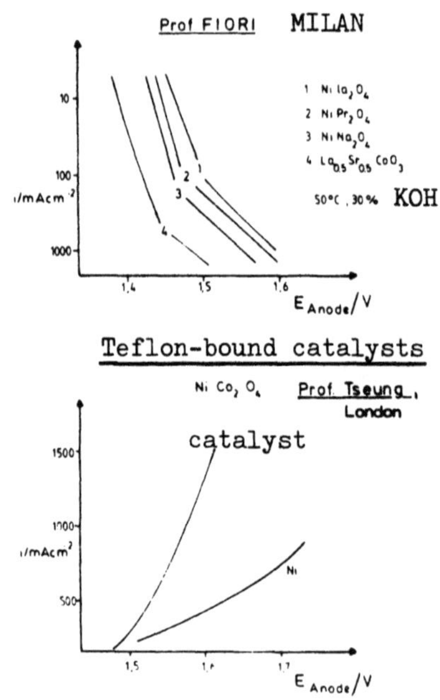

Figure 4   Examples of anodic electrocatalysis

EEC programme - Advanced water electrolysis

Starting-point: conventional electrolysis

30 wt.% KOH, 80-90°C, activated cathodes

A   Improvement of individual components

A.1 Anode activation using mixed oxides containing cobalt (perovskites, spinels, etc)

A.2 Cathode activation using finely divided deposits of nickel, cobalt or iron

A.3 Very thin corrosion-resistant diaphragms and membranes with very low resistance per unit area:

a) PTFE-bound K-hexatitanate
b) inorganic ion exchangers
c) metal-supported porous oxide ceramics
d) polymer-reinforced asbestos
e) new organic ion exchange membranes

A.4 Modification of cell design by using thin-layer techniques

<u>B</u> Raising the working temperature of conventional water electrolysis

<u>C</u> Improvement of the Solid Polymer Electrolyte (SPE) concept

<u>D</u> High-temperature electrolysis of water

SCK / CEM Mol

<u>Membrane</u>: polyantimonic acid cation exchanger on a plastic support

Anode: $NiCo_2O_4$ Cathode: nickel

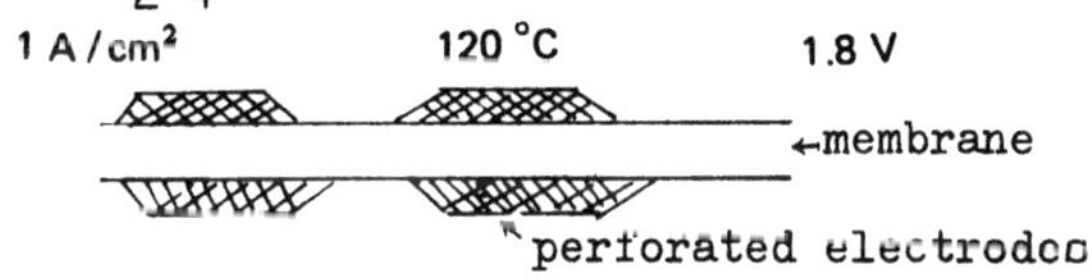

<u>Diaphragms</u>

Giuffré Politecnico Milano:

Copolymers: vinylpyridine - divinylbenzene

Polymer-reinforced asbestos membranes

<u>SPE concept</u>

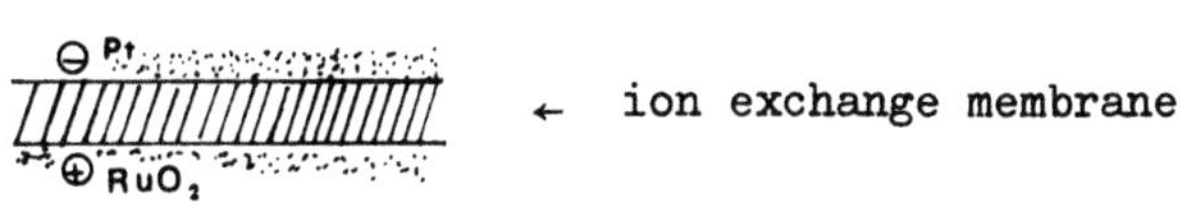

Replacement of Nafion by inexpensive ion exchangers
Replacement of anodes and cathodes containing noble

metals by other types of electrode

<u>High-temperature electrolysis</u>

HOT-ELLY 850-950°C (Dornier system)

or perovskite

300-500 µm thickness ⟶ 50-100 µm

a) cylindrical electrodes (Dornier)
b) conical plates (Citroen)
c) plates with a honeycomb structure (CRN, Grenoble)

Examples illustrating R&D projects carried out by the EEC:

Investigations and results from the Technische Hochschule, Darmstadt
(Institut für Chemische Technologie)
(J. Fischer, H. Hofmann, G. Luft and H. Wendt: Fundamental Investigations and Electrochemical Engineering Aspects concerning an Advanced Concept for Alkaline Water Electrolysis, AIChE J. Vol 26, pp 794-801)

1. Current-voltage curves of electrolytic cells with unactivated and activated anodes (Figure 5)

2. Representation of the sandwich cell configuration with perforated electrodes to minimize the masking effect of bubbles. Based on oxide ceramic diaphragms supported on nickel mesh (Figure 6)

3. Cell voltages for water electrolysis at 1 A/cm^2 and 50 bars with different cell geometries and types of anode activation (Figure 7)

4. Principle of a pilot plant (five 1-kW units) for long-term testing of electrolyser components (Figure 8)

5. Cost estimates for the electrolytic production of hydrogen (Figure 9)

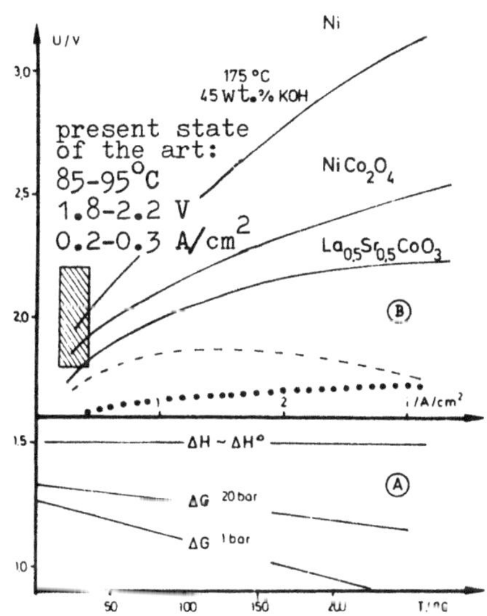

Figure 5 Demonstration of anodic catalysis and the masking effect of bubbles:
The upper part of the figure shows current-voltage curves measured with unactivated and activated plane nickel anodes and roughened nickel cathodes.
- - - -: calculated IR correction for activation by La, Sr, Co perovskite;
.......: anodes activated with La, Sr, Co perovskite using the optimum cell configuration

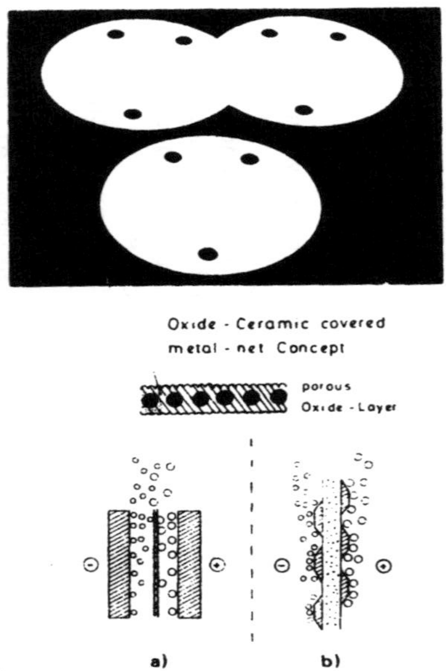

Figure 6 Cell configurations:
a) conventional
b) sandwich

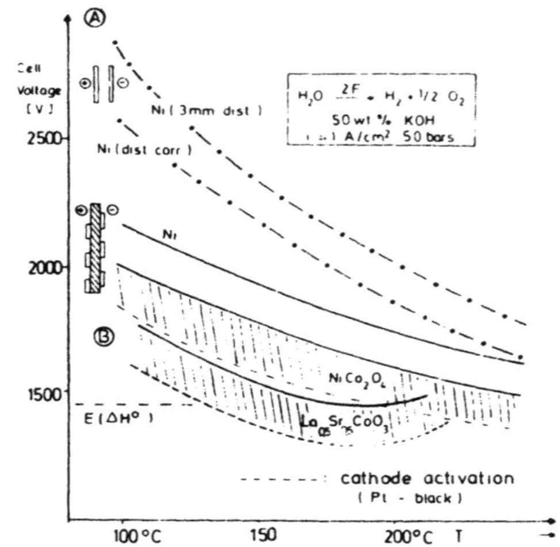

Figure 7 Temperature dependence of the cell voltage in water electrolysis for different cell configurations and types of electrode activation

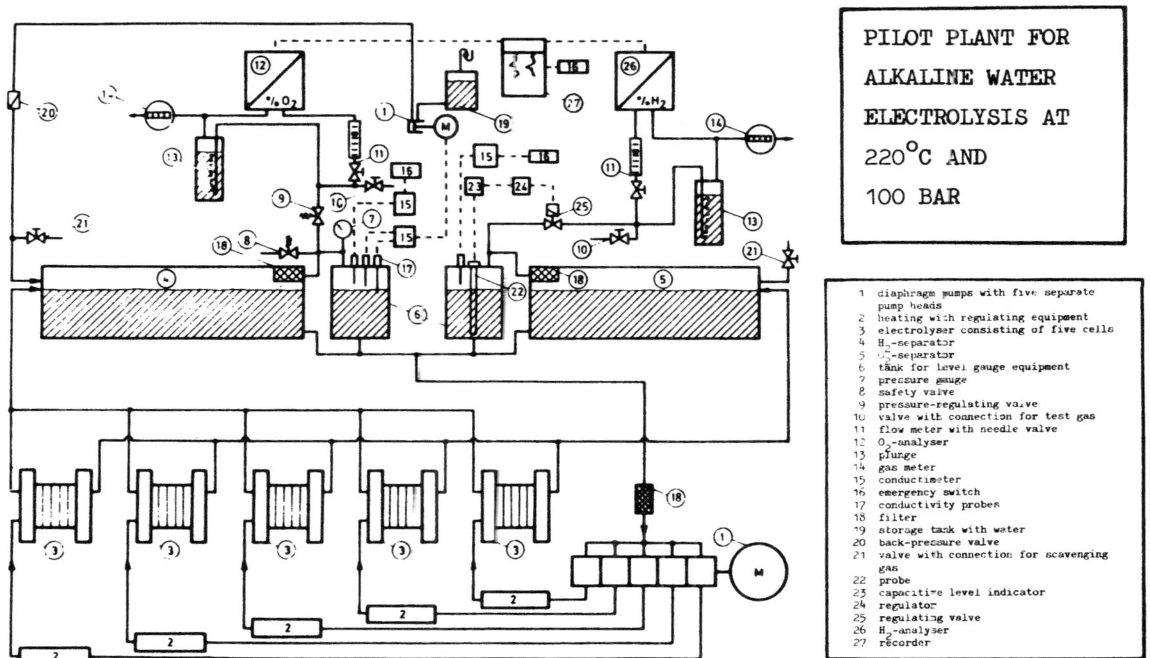

Figure 8 Principle of the design and operation of a pilot plant (5 x 1 kW) for the long-term investigation of newly developed electrolyser components

Electrolyser	Lurgi	Electrolyser with filter-press arrangement (Lurgi) incorporating newly developed cell components and sandwich configuration		
		Version 1*	Version 2**	Version 3***
U_{cell}, V	1.9	2.0	1.65	1.80
Current density A/cm^2	0.2-0.3	1.0	1.0	1.5
Working temperature °C	90	90	170	170
Working pressure bar	30	30	60	60
Capital costs US$/$Nm^3$ H_2	0.0827	0.0497	0.0620	0.04152
Energy costs US$/$Nm^3$ H_2	0.1372	0.1428	0.1181	0.1287
Maintenance US$/$Nm^3$ H_2	0.0140	0.0140	0.0140	0.0140
Final price US$/$Nm^3$ H_2	0.234	0.212	0.194	0.1842

Figure 9 Estimated cost of producing hydrogen electrolytically by different methods (compared with the Lurgi process)

* conventional steel construction; 1 A/cm^2
** steel construction with 1 mm nickel plating; 1 A/cm^2
*** version 3 identical with version 2; 1.5 A/cm^2
Cost of energy US$ 0.026/kWh; 100% utilization. Writing-off within 5 years.

Thermochemical processes for hydrogen production

by G.H. Schütz[1]

1. Introduction

For the last 10 years or so the development of a new method for producing hydrogen has been followed with considerable interest. This is the production of hydrogen by means of thermochemical reactions, which constitutes an alternative to the electrolysis of water.

The present paper gives a brief description of this method, the requirements which it must satisfy, and the chemical reactions which are involved. This is followed by a review of the processes which are being most actively investigated at present. Finally, an indication is given of the conclusions that can be drawn from the work carried out so far in this field.

2. Methods for the production of hydrogen

Methods for producing hydrogen can be divided in principle into two groups. The first group covers processes in which water is the sole raw material, and is decomposed into hydrogen and oxygen by supplying energy. In the second group other raw materials are used in addition to water, giving hydrogen together with other products. The second group covers all the methods for producing hydrogen which are of importance at present, such as steam reforming, for example, in which hydrocarbons (and sometimes oxygen) are used in addition to water and the products are hydrogen to-

1. Dr G. Schütz, Commission of the European Communities, Ispra Joint Research Centre, Ispra/Varese, Italy

gether with carbon monoxide or carbon dioxide. Systems in which sulphur or sulphur compounds are used as additional raw materials have also recently received attention in connection with desulphurization processes (1,2,3). We shall be concerned exclusively with the first group.

The rapid decline in fuel resources, especially of liquid fossil fuels, and the possibility of introducing hydrogen as a secondary energy vector (4,5,6), have in recent years directed much attention to the production of hydrogen from water as the sole raw material.

Water can of course be decomposed in a single step by supplying electrical energy. The electrolysis of water is an established industrial process, but it is competitive nowadays only when cheap hydro-power is available. Since the thermal energy must first be converted to electricity, and overvoltages occur at the electrodes, the attainable overall efficiency is at present only about 25%. However, water can also be decomposed by other forms of energy, of which we shall here consider only thermal energy.

3. The thermochemical production of hydrogen

Direct thermal decomposition requires temperatures above 2500 K, and is therefore not industrially feasible in the foreseeable future. However, it has been found that water can be decomposed at far lower temperatures, more readily obtained industrially, by means of a series of several chemical reactions. The chemical substances used for this pass through a cycle, and such processes are therefore called thermochemical cyclic processes. They constitute systems in which only water and energy are consumed, while hydrogen and oxygen are produced together with waste heat, ie thermal energy is converted into chemical energy.

Funk and Reinstrom (7) were the first to suggest that it might be possible to use a multi-stage process to produce hydrogen from water alone with a thermal efficiency greater than that of electrolysis. Experimental investigations began a few years later, first at the Ispra Joint Research Centre and soon afterwards at several laboratories in the USA, West Germany, France and Japan.

The high-temperature nuclear reactor appears to be the most suitable primary energy source for producing hydrogen on a large scale. The high-grade process heat that can be obtained from its cooling circuit served from the very beginning to stimulate research into the thermochemical production of hydrogen. The AVR (Arbeitsgemeinschaft Versuchsreaktor GmbH) experimental nuclear power plant at Jülich has been operated since 1974 with an average gas exit temperature of 1225 K (8). Temperatures about 100 K higher could probably be obtained with further development. These values are taken as a guide for the use of nuclear energy for thermochemical cyclic processes and suggest a maximum process temperature of 1000-1200 K.

As well as nuclear energy, the use of solar and even fusion energy has been suggested (9,10). The solar power plants so far constructed give temperatures of 775-975 K, but far higher temperatures, even above 1300 K, are possible. However, at high temperatures capital costs rise at a disproportionate rate, and there are increased problems connected with constructional materials and loss of heat by radiation. This also applies to the fusion reactor, which should give heat at 1600-1700 K.

Thermochemical cyclic processes can be classified from different points of view, eg according to the number of chemical reactions, the maximum temperature, etc. It is customary to divide them into families, within which a number of processes employ the same elements or compounds but in different combinations. Examples are the iron-chlorine family, intensively studied a few years ago (11,12), and the sulphur-halogen family which has replaced it.

It is also important to distinguish between purely

thermochemical processes, which essentially require only heat to bring about the reactions, and hybrid processes, which employ another form of energy (usually electrical) in addition to heat. Hybrid processes with one electrochemical stage are of interest only if - as at the present time - the cost of electricity is high, and if electrolysis takes place at considerably lower cell voltages than the electrolysis of water. Roughly speaking, the savings on electrical energy must more than compensate for the capital and energy costs of the additional chemical plant required. The advantages of hybrid processes over purely thermochemical ones lie mainly in the fact that they may make it possible to use reaction sequences which are difficult to carry out purely thermochemically or which require more stages, and that they generally reduce the effort required to separate the products arising in the process. The hybrid processes which have been farthest developed carry out the electrochemical stage at a low temperature, thus giving pure hydrogen immediately. Low temperatures also help to reduce the problems connected with constructional materials which often arise in thermochemical cyclic processes.

4. Requirements to be satisfied by a thermochemical cyclic process

A thermochemical cyclic process is subject to numerous restrictions. In the first place its overall thermal efficiency must be appreciably higher than that of the electrolysis of water. This efficiency can be defined in various ways, depending on the viewpoint adopted. To enable comparisons to be made between individual processes, an IEA working party has therefore laid down that the thermal efficiency η should be defined as the ratio between the maximum calorific value of the hydrogen and Q, the total amount of energy fed into the system in no matter what form, expressed in thermal equivalents (kJ):

$$\eta = \frac{-\Delta H^{o}(H_2O_{fl})}{Q} = \frac{285.9}{Q} .$$

The efficiency can also be calculated or estimated from existing thermodynamic data according to established rules. However, in practice one must generally reckon with a considerably lower value, since the reactions do not follow the ideal course assumed in the calculation, and the extra energy needed for separating and transporting the chemical products must be taken into account. Favourable thermodynamics, ie $\Delta G \approx 0$ or slightly negative for the individual reactions, is not sufficient alone to ensure that the process is industrially and economically practicable: this demands above all that there should be few stages and that the reaction should be sufficiently fast. However, it follows from thermodynamic considerations that below the temperature range of about 1000-1200 K, in which the reactions must occur, at least three stages are required to decompose water thermochemically (7,13). The greater the number of stages, the lower will be the efficiency. Only in hybrid processes are two stages possible. Fast reactions mean short circulation times, and hence lower costs for plant and the chemicals used. The latter should as far as possible be easy to handle, not too toxic, and available in sufficient quantities.

The products of the individual reactions should be obtained in satisfactory yields, so as to minimize the effort involved in separating them. Dilute solutions are to be avoided if possible, since too much heat is lost in working them up. Homogeneous gas reactions are preferred in general. They must, however, take place rapidly, since otherwise the advantage of a process which is simple to carry out may be partly or wholly lost in the working up of the products. Side reactions should also be excluded as far as possible. Further, any contamination of the products of one stage must not affect the succeeding reaction. These and other factors which must be taken into account in seeking for a thermochemical cycle have naturally been advanced by many critics, but they have not discouraged others from proposing numerous processes or indeed from investigating them experimentally.

5. Fundamental types of reaction for thermochemical cyclic processes

The fundamental types of reaction for thermochemical cyclic processes can be divided into those which form hydrogen, those which form oxygen, those which decompose water, or hydrolysis reactions, and those which regenerate intermediate products. Since the decomposition of water requires both oxidation and reduction, multivalent elements or compounds are used for this purpose.

1) Formation of hydrogen

These reactions usually occur exothermically at low or moderate temperatures, either

a) by the reaction of a metal with an acid, mostly a hydrogen halide, or

b) by using a hydrogen halide to oxidize a metallic halide to a halide of higher valency, or

c) by the thermal or electrolytic decomposition of hydrides, eg:

a) $2\ Cu + 2\ HCl \rightarrow 2\ CuCl + H_2$
b) $2\ VCl_2 + 2\ HCl \rightarrow 2\ VCl_3 + H_2$
c) $2\ HI \rightleftarrows I_2 + H_2$.

Hydrogen iodide can be decomposed thermally or electrolytically in a single step, while hydrogen chloride and hydrogen bromide can be decomposed either electrolytically in one stage or thermally in two stages. The thermal decomposition of hydrogen selenide or arsenide gives much higher yields of hydrogen than are obtained from hydrogen iodide.

2) Formation of oxygen

This generally occurs at high temperatures by the decomposition of

a) the oxide of a metal, or
b) the oxide of a non-metal, or
c) oxyacids or their salts.

The evolution of oxygen from metallic oxides is not at

present of importance, since (with the exception of mercuric oxide) the necessary temperatures cannot be obtained from the high-temperature nuclear reactor. On the other hand great interest attaches to the thermal decomposition of sulphur trioxide, metallic sulphates and sulphuric acid. The last-mentioned takes place according to the equation:

$$H_2SO_4 \rightarrow H_2O + SO_2 + 1/2\ O_2.$$

In many cyclic processes the decomposition of sulphuric acid constitutes the high-temperature stage. It can be carried out easily in the laboratory, but on an industrial scale there are serious problems connected with constructional materials.

3) Decomposition of water

Water can be brought into the cycle and caused to react over a wide range of temperatures. Special interest attaches to the reverse Deacon reaction, which occurs at high temperatures with the evolution of oxygen:

$$Cl_2 + H_2O \rightarrow 2\ HCl + 1/2\ O_2$$

and to the reaction of water with bromine or iodine in the presence of sulphur dioxide (Bunsen reaction), which occurs at low temperatures:

$$X_2 + 2\ H_2O + SO_2 \rightarrow 2\ HX + H_2SO_4.$$

6. Thermochemical cyclic processes at present under investigation

Numerous candidate cycles have been suggested during the past 10 years. The technical literature (14,15) contains a collection of 201 processes in all. In addition several thousand have been discovered by means of computer programs (16,17,18,19). However, the application of strict criteria leaves only a few which merit further investigation. A large proportion of the suggestions have not been investigated experimentally. Others have been abandoned when it was shown by a more thorough study of the literature, and also by experiments, that one or more of the reactions could be carried out in practice only with great difficulty, or not at all. Here, we shall consider only processes which are being worked on at the present time. Since there are more than 30 of these we shall take into account only those which are at an advanced stage of development, or are of special importance. Three cyclic processes should be mentioned particularly, for which demonstration models are already available, constructed mainly of glass or quartz and giving a continuous production of about 100 l hydrogen per hour. These are the Westinghouse sulphur cycle, the Ispra Mark 13 bromine-sulphur cycle and the General Atomic Co. iodine-sulphur process. Of these, the first two are hybrid processes (20,21,22).

6.1 Westinghouse sulphur cycle

$$SO_2 + 2\,H_2O \xrightarrow{298\text{-}373\ K} H_2SO_4 + H_2 \qquad \text{electrolysis}$$

$$H_2SO_4 \xrightarrow{925-1125\ K} H_2O + SO_2 + 1/2\ O_2 \qquad \text{thermolysis}$$

In this process sulphur dioxide in sulphuric acid solution is oxidized electrochemically to sulphuric acid, hydrogen being formed at the cathode. The sulphuric acid is highly concentrated, decomposed at a high temperature, and the sulphur dioxide and water formed are used again in the process.

The electrochemical stage was proposed by Juda and Moulton as long ago as the 1960s (23) for the simultaneous production of sulphuric acid and hydrogen. The cycle was first investigated by Bowman and Onstott (24) and was subsequently studied intensively by the Westinghouse Electric Corporation, USA (25,26). Many investigators have been attracted by the fact that the process consists of only two steps, that the standard potential for the oxidation of sulphur dioxide to sulphuric acid is low (0.17 V), that the chemicals used are cheap, and that only liquids and gases are involved. For this reason several laboratories, eg the Ispra Joint Research Centre (Mark 11 cycle) and the Kernforschungsanlage Jülich, are at present working on the same process. However, although at first sight the process appears to be intriguingly simple, there are serious difficulties in carrying it out in practice. The electrochemical stage functions best in dilute sulphuric acid - an optimum of 30 wt.% has recently been cited (27) - whereas highly concentrated acid is most favourable for the high-temperature stage. If, however, higher concentrations are used in the electrolytic cell both the resistance of the electrolyte and the anode potential increase considerably. Also, from a concentration of about 70-75% onwards there is reduction of the acid at the cathode to sulphur and hydrogen sulphide. All this makes it necessary to optimize the acid concentration for the two stages and to use considerable amounts of heat in producing highly concentrated acid.

There is a further problem connected with the electrolytic cell. In order to avoid losses of potential and of hydrogen at the cathode the cathode compartment must be kept free from sulphur dioxide. This can be achieved either by a higher static pressure of the electrolyte on the cathode side (28) or by a specially constructed cell containing two cation exchange membranes, between which there flows an auxiliary electrolyte (29). The electrodes are made of porous graphite, graphite gauze or graphite felt. Both the anodic and the cathodic reactions must be catalysed. It has not yet proved possible to attain the low voltage of about 0.45 V at 2 kA/m^2 that was initially anticipated. Nevertheless, recent results obtained with small laboratory cells are quite interesting. In one experiment lasting for 173 hours the cell voltage was 675 mV at 100 mA/cm^2 and 323 K (27), while in another instance with the addition to the anolyte of small amounts of hydrogen iodide as a catalyst the voltage was 565 mV at 200 mA/cm^2 and 363 K (29). In both cases 50% sulphuric acid was used.

Detailed balance sheets and cost analyses have been carried out for this process (25). Its overall efficiency is estimated at about 45%. In order to avoid the serious problems with constructional materials for the decomposition of sulphuric acid and the need to concentrate the acid to a high degree, thus making it possible to utilize the lower cell voltages obtainable with more dilute sulphuric acid, the Los Alamos Scientific Laboratories have suggested the following four-stage process (30):

$$2\,H_2O + SO_2 \xrightarrow{350\ K} H_2SO_4 + H_2 \qquad \text{electrolysis}$$
$$H_2SO_4 + 1/3\,Bi_2O_3 \xrightarrow{350\ K} 1/3\,Bi_2O_3 \cdot 3\,SO_3 + H_2O$$
$$1/3\,Bi_2O_3 \cdot 3\,SO_3 \xrightarrow{900-1200\ K} 1/3\,Bi_2O_3 + SO_3$$
$$SO_3 \xrightarrow{900-1500\ K} SO_2 + 1/2\,O_2.$$

The formulation given here shows as an example a cycle with normal bismuth sulphate. There are several alternatives with various oxysulphates which permit the use of 3-52.7% sulphuric acid and which also have a lower heat of decomposition. In the suggested process a metallic oxide or oxysulphate is added to the dilute sulphuric acid outside the cell to precipitate a solid sulphate, which is decomposed at a high temperature into oxygen, sulphur dioxide and a metallic oxide or oxysulphate. This has the advantage of entailing less corrosion by the solid sulphates and the gaseous products in the decomposition reactor: on the other hand, the mass flow is greater, and solid substances have to be filtered, dried, transported and caused to react. It is not yet clear how much heat would be saved in practice by the elimination of the need to concentrate the sulphuric acid, since the product which is filtered off contains water of crystallization (eg 3 moles of water for $Bi_2O_3 \cdot 3\,SO_3$) in addition to the usual associated moisture.

For the electrochemical stage it is important to investigate whether the Bi^{3+} ions present in the cell cycle might affect the catalysis and thus increase the cell voltage. This process is being studied with special reference to the use of solar energy or even fusion energy.

An overall efficiency of 50.8% has been calculated for an upper temperature of 1500 K.

6.2 Ispra Mark 13 cycle

$$2\,HBr \xrightarrow{353-473\,K} H_2 + Br_2 \qquad \text{electrolysis}$$

$$Br_2 + SO_2 + 2\,H_2O \xrightarrow{293-373\,K} 2\,HBr + H_2SO_4aq$$

$$H_2SO_4 \xrightarrow{925-1125\,K} H_2O + SO_2 + 1/2\,O_2 \qquad \text{thermolysis}$$

In the first stage concentrated hydrobromic acid is electrolysed at about 373 K, giving hydrogen and bromine. The latter is separated by distillation, and in the second stage reacts with sulphur dioxide in an aqueous solution of hydrogen bromide and sulphuric acid. One thus obtains gaseous hydrogen bromide and fairly concentrated sulphuric acid (about 75-80%), which can then be decomposed at high temperatures, either directly or after further concentration. It is well known that hydrogen bromide cannot be appreciably decomposed thermally in a single stage below 1300 K. As in the case of water, direct decomposition is possible only with the aid of electrical energy. If one compares the standard cell potentials for the two electrolyses, the small difference is rather discouraging (1.23 - 1.07 = 0.16 V). However, for hydrobromic acid it is possible to lower the reversible cell voltage markedly (by several 100 mV), notably by increasing the concentration of hydrogen bromide. Industrial realization of the electrolysis of hydrobromic acid has been well investigated (31,32, 33,34).

Graphite is used as the electrode material. The overvoltage for the discharge of hydrogen can be kept low by the addition of small amounts of noble metal salts. No catalyst is needed at the anode. The overvoltage for the discharge of bromine is very small (in contrast to that for chlorine), and can be neglected in practice.

At 373 K cell voltages of 0.75-0.8 V at 2 kA/m^2 are obtained. Long-term experiments have shown that there are no obstacles to the industrial realization of the electrolysis of hydrobromic acid. The second stage raises no particular problems on a laboratory scale.

At the present stage it is difficult to make a comparison with the Westinghouse process. The somewhat lower cell voltage in the Westinghouse sulphur process is counterbalanced by the advantage of the higher concentration of sulphuric acid obtained in the Mark 13 process. It should also be mentioned that the incorporation of an additional element - bromine - in the Mark 13 process may also be advantageous, since it offers more possibilities in the industrial operation of the process. For example, the cycle can be modified so that no free bromine is formed. Purification of the oxygen from sulphur dioxide or separation of the mixture of oxygen and sulphur dioxide can be achieved by the Bunsen reaction. Electrolysis can be carried out with either one or two cell cycles. A further reduction of the cell voltage can be obtained by carrying it out under pressure. Without departing from the conventional pressure range, the use of a precipitation technique for the sulphuric acid makes it possible to obtain at 373 K a cell voltage of 0.7 V at about 5 kA/m^2 (35).

These considerations show that further work is necessary to determine which is the optimum process.

6.3 General Atomic Co. iodine-sulphur cycle

Among the purely thermochemical cyclic processes, those belonging to the iodine-sulphur family are of most interest at the present time. The following three-stage process has been developed by the General Atomic Co. in particular (36):

$$2\,HI \xrightarrow{393-800\,K} H_2 + I_2$$

$$I_2 + SO_2 + 2\,H_2O \xrightarrow{370\,K} H_2SO_4 + 2\,HI$$

$$H_2SO_4 \xrightarrow{1150\,K} H_2O + SO_2 + 1/2\,O_2.$$

The main difficulty in this process lies in the fact that, if side reactions are to be avoided, the two acids can be obtained only in dilute solution in the second stage, and are difficult to separate even on a laboratory scale. Separation by distillation is impossible without decomposition. The General Atomic Co. has developed a continuous process in which an excess of iodine is used to effect separation into approximately 56% sulphuric acid and hydriodic acid rich in iodine. After phosphoric acid has been added to the latter, treatment at a high temperature yields liquid iodine and gaseous hydrogen iodide. The further working up of the solutions consumes a considerable amount of heat. Nevertheless, an overall efficiency of 47% is attained.

The difficulties associated with this process have led in recent years to numerous investigations and suggestions for modifying the process. For instance, Dokya and his collaborators have attempted to carry out the second stage electrochemically (37). In this way the two acids are formed separately. On short-circuiting the cell the following electrode reactions occur:

cathode: $I_2 + 2H^+ + 2e^- \rightarrow 2HI$

anode: $SO_2 + 2H_2O \rightarrow H_2SO_4 + 2H^+ + 2e^-$.

The anode and cathode compartments are separated by an ion exchange membrane. However, the current densities so far obtained have been too low, and the properties of both the ion exchange membranes used and the electrode materials have been unsatisfactory.

The Japan National Chemical Laboratory for Industry is investigating the reduction of iodine in the presence of methanol (38), which gives sulphuric acid together with readily volatile methyl iodide according to the following equation:

$$I_2 + SO_2 + 2CH_3OH \xrightarrow{H_2O} H_2SO_4 + 2CH_3I.$$

The sulphuric acid is decomposed thermally, as in the other processes, and the methyl iodide is converted into hydrogen, methanol and iodine to complete the cycle. There are several possibilities for this last conversion, which have also been investigated, notably in Japan, from the viewpoint of industrial realization.

For instance, the following route leads to methanol and hydrogen via the hydrogenation of methyl iodide:

$$2\,CH_3I + H_2 \rightarrow 2\,CH_4 + I_2$$
$$2\,CH_4 + 2\,H_2O \rightarrow 2\,CO + 6\,H_2$$
$$4\,H_2 + 2\,CO \rightarrow 2\,CH_3OH.$$

Many modifications have also been proposed for the thermal decomposition of hydrogen iodide. Hydrogen iodide can of course be readily decomposed thermally, but the yield of hydrogen is relatively low on account of the unfavourable position of equilibrium, which varies little with temperature: this leads to large mass flows and large quantities of heat to be exchanged. The General Atomic Co. is therefore investigating thermal decomposition in the liquid phase at considerably lower temperatures with the addition of catalysts (39). Electrochemical decomposition is also being considered. Cell voltages of 0.25 V at 2 kA/m^2 and 373 K were obtained in laboratory experiments (35).

The Japan Atomic Energy Research Institute's nickel-iodine-sulphur (NIS) process decomposes the hydrogen iodide and separates the hydriodic acid from the sulphuric acid by using nickel salts (40):

$$2\,H_2O + I_2 + SO_2 + 2\,Ni \xrightarrow{298-370\,K} NiI_2 + NiSO_4 + 2\,H_2$$
$$NiI_2 \xrightarrow{875-975\,K} Ni + I_2$$
$$NiSO_4 \xrightarrow{1125\,K} NiO + SO_2 + 1/2\,O_2$$
$$NiO + H_2 \xrightarrow{675-875\,K} Ni + H_2O.$$

This process is characterized by the absence of water and of hydrogen from both high-temperature stages - the decomposition of nickel iodide and of the sulphate. The two salts formed in the first step are separated from one another by solvent extraction. The evaporation of the solvent and the removal of water of crystallization are the main reasons for the low overall efficiency of about 34%.

Mention should also be made of attempts to decompose

hydrogen bromide thermally in several stages, instead of electrolytically. At the Government Industrial Research Institute in Osaka, Japan, the following cyclic process is being investigated, among others (41):

$$3\,FeBr_2 + 4\,H_2O \xrightarrow{893-1273\,K} Fe_3O_4 + 6\,HBr + H_2$$

$$Fe_3O_4 + 8\,HBr \xrightarrow{473-573\,K} 3\,FeBr_2 + 4\,H_2O + Br_2$$

$$Br_2 + 2\,H_2O + SO_2 \xrightarrow{293-373\,K} H_2SO_4 + 2\,HBr$$

$$H_2SO_4 \xrightarrow{1150\,K} H_2O + SO_2 + 1/2\,O_2.$$

However, experiments of this kind have not so far led to any process which is superior to the electrolysis of hydrobromic acid.

At the Rheinisch-Westfälische Technische Hochschule in Aachen, in addition to iodine-sulphur processes the following vanadium-chlorine cyclic process is being investigated (42):

$$2\,VCl_2 + 2\,HCl \xrightarrow{333\,K} 2\,VCl_3 + H_2$$

$$2\,VCl_3 \xrightarrow{960\,K} VCl_2 + VCl_4$$

$$VCl_4 \xrightarrow{425\,K} VCl_2 + Cl_2$$

$$Cl_2 + H_2O \xrightarrow{1200\,K} 2\,HCl + 1/2\,O_2.$$

This process, plus a similar three-stage process with tantalum chloride, were the first thermochemical processes to be proposed, but in view of their low calculated efficiency they were not at first tested experimentally (43). Knoche and his collaborators have succeeded in improving the efficiency from 18% to about 40%.

At the Lawrence Livermore Laboratory, USA, the following process is being investigated (44):

$$H_2Se \xrightarrow{750\ K} Se + H_2$$

$$2\ ZnO + Se + SO_2 \xrightarrow{800\ K} ZnSe + ZnSO_4$$

$$ZnSe + 2\ HCl \xrightarrow{350\ K} ZnCl_2 + H_2Se$$

$$ZnCl_2 + H_2O \xrightarrow{900\ K} ZnO + 2\ HCl$$

$$ZnSO_4 \xrightarrow{1200\ K} ZnO + SO_2 + 1/2\ O_2.$$

While the decomposition of hydrogen selenide takes place rapidly and with a high yield of hydrogen (about 60%), the reaction of zinc oxide with selenium and sulphur dioxide has so far proved to be too slow.

Investigations are at present concentrated on the thermal decomposition of zinc sulphate, which could possibly also be used in other cycles to replace the decomposition of sulphuric acid (45).

A number of hybrid processes have been proposed by Behr. Their distinguishing feature is that the electrochemical step is carried out in fused salts at temperatures of 600-950 K (46,47). The following process is being investigated, among others:

$$H_2O + (2\ Li) \xrightarrow[NaOH]{600-800\ K} (2\ LiH) + 1/2\ O_2 \quad \text{electrolysis}$$

$$2\ LiH \xrightarrow{1000-1200\ K} 2\ Li + H_2 \quad \text{thermolysis.}$$

In practice this process amounts to the electrolysis of water in fused NaOH. Hydrogen atoms diffuse through a membrane electrode (cathode) and combine with lithium to form lithium hydride, which is decomposed at a high temperature. The main difficulty is to find materials for the membrane and electrodes which are suitable for the desired temperature range.

7. Conclusions

Experimental work on thermochemical cyclic processes started about 10 years ago. This field is therefore only in its initial stages and is still insufficiently investigated. Although these new processes are receiving attention world-wide, the resources applied to them are still quite modest. Both theoretical considerations and the experimental work carried out so far show that it is difficult to fulfil simultaneously the numerous conditions which must be satisfied by a thermochemical cyclic process. Only a relatively small number of the many suggestions merit more detailed investigation. However, it cannot be asserted that these are the most favourable ones. It is quite possible that still better combinations of reactions will be discovered in the future. Investigations are hindered by the absence or insufficient accuracy of thermodynamic data, and above all by the lack of sufficiently accurate kinetic information, which is rarely available.

Some of the important targets aimed at have already been reached. For instance, cycles carried out on a laboratory scale have shown that such processes could be carried out industrially in the required temperature range without the formation of by-products. Those processes which have been most developed technologically exhibit realistic overall efficiencies of more than 40%, and in this respect constitute a marked advance compared with the conventional electrolysis of water. There are also some combinations of reactions with an overall efficiency of 50-55%, though these await further experimental investigation. In comparing the two competing procedures one must also take into account the improvements which are to be expected in the electrolysis of water. In recent years interesting results in this direction have been obtained in laboratory experiments, but it remains to be seen whether these can be realized in the long term under operational conditions. The realization of thermochemical cyclic processes does not really require any special technological breakthroughs - merely the improvement of existing materials and engineering practice. Many stages of the process, for example electrochemical stages, could certainly be carried out on an industrial scale.

There is still much uncertainty about the capital costs of thermochemical cyclic processes. Sufficiently precise estimates must await the construction of pilot plants.

As regards the high-temperature stage, especially the decomposition of sulphuric acid, further investigation is needed to determine whether the problems involving constructional materials and coupling to the heat source can be solved satisfactorily, or whether it would be better to replace this stage by something different, eg the decomposition of a sulphate. A pilot plant for the decomposition of 45 kg of sulphuric acid per hour by means of hot air is at present under construction (48).

Further developments in the field of high-temperature nuclear reactors are also important for the future of thermochemical hydrogen. A slowing down in this field could delay or even prevent the development of pilot or prototype plants using thermochemical cyclic processes. This could lead to a shift of research activity towards other methods, such as the photochemical production of hydrogen, or a trend towards the two other high-temperature sources of heat, solar and fusion energy, as can be noticed already in the United States.

At the present time neither water electrolysis nor thermochemical processes can compete with current methods of producing hydrogen. In the present state of knowledge there is no justification for preferring either of these methods to the other.

It must not be forgotten that in the future thermochemical hydrogen could make a valuable contribution to an efficient and cleaner energy system.

References

(1) G.H. Parker, G.H. Farbman and W.A. Summers: An Early Application of the Sulfur Cycle Hydrogen Production Process, Proc. 3rd WHEC, Tokyo, Japan, 23-26 June 1980

(2) G.H. Schütz: Processes for Combined Production of Hydrogen and Other Chemical Products, Proc. 3rd WHEC, Tokyo, Japan, 23-26 June 1980, p 463

(3) D. van Velzen, H. Langenkamp and G. de Beni: Process for the Removal of SO_2 from Waste Gases, Producing Hydrogen and Sulfuric Acid, UK Patent Application No 7910315, 23 March 1979

(4) G. de Beni and C. Marchetti: Hydrogen, Key to the Energy Market, Euro Spectra, 9, 46 (1970)

(5) R.L. Savage et al: A Hydrogen Energy Carrier, NASA Pub. NGT 44-005-114, September 1973, Houston, Texas (USA)

(6) D.P. Gregory: A Hydrogen-Energy System, American Gas Association, 1973

(7) J.E. Funk and R.M. Reinstrom: Energy Requirements in the Production of Hydrogen from Water, Ind.Eng. Chem.Proc.Des.a.Dev., 5, 336 (1966)

(8) H. Barnet: Nutzung der nuklearen Prozesswärme für die Gewinnung von Wasserstoff (Utilization of nuclear process heat for the production of hydrogen), see elsewhere in this publication

(9) B.J. Eastlund and W.C. Gough: Generation of Hydrogen by Ultraviolet Light produced by the Fusion Torch, paper presented at the 163rd ACS Meeting, Boston, USA (1972)

(10) M.G. Bowman: Interfacing Primary Heat Sources and Cycles for Thermochemical Hydrogen Production, Proc. 3rd WHEC, Tokyo, Japan, 23-26 June 1980, p 335

(11) D. van Velzen and H. Langenkamp: Problems around Fe-Cl Cycles, Int. J. Hydr.Energy, 3, 419 (1978)

(12) K.F. Knoche: Stand der Arbeiten zur Wasserstoff-Erzeugung mit nuklearer Prozesswärme (Status of work on hydrogen production using nuclear process heat), Chem.-Ing.-Tech., 49, 239 (1977)

(13) B.M. Abraham and F. Schreiner: General Principles underlying Chemical Cycles which Thermally Decompose Water into the Elements, Ind.-Eng.-Chem. Fundam., 13, 305 (1974)

(14) C.E. Bamberger and D.M. Richardson: Hydrogen Production from Water by Thermochemical Cycles, Cryogenics, 16, 197 (1976)
(15) C.E. Bamberger: Hydrogen Production from Water by Thermochemical Cycles; a 1977 Update, Cryogenics, 18, 170 (1978)
(16) J.L. Russel and J.T. Porter: A Search for Thermochemical Water-splitting Cycles, THEME Conf. Proc. Miami Beach, Florida, USA (1974), pp 11-49
(17) K.F. Knoche, H. Cremer and G. Steinborn: A Thermochemical Process for Hydrogen Production, Proc. Hydrogen Energy Fundamentals Symposium (1975)
(18) H. Kameyama, K. Yoshida and D. Kunii: 39th Annual Meeting of the Chem.Eng.Soc., Japan (1974)
(19) J.E. Funk, W.L. Conger and R.H. Carty: THEME Conf.Proc., Miami Beach, Florida, USA (1974), pp. 11.1-11.8
(20) G.H. Parker and P.W.T. Lu: Laboratory Model and Electrolyser Development for the Sulfur Cycle Hydrogen Production Process, Proc. 14th IECEC Meeting, Boston, Mass, USA (1979)
(21) D. van Velzen, H. Langenkamp, G.H. Schütz, D.R. Lalonde, J. Flamm and P.J. Fiebelmann: Development, Design and Operation of a Continuous Laboratory-Scale Plant for Hydrogen Production by the Mark-13 Cycle, Proc. 2nd WHEC, Zürich, 21-24 August 1978, p 649
(22) G. Besenbruch et al: Development of a Sulfur-Iodine Thermochemical Water-splitting Cycle for Hydrogen Production, Proc. 14th IECEC Meeting, Boston, Mass, USA (1979)
(23) W. Juda and D.M. Moulton: Effect of Low-Cost Electrolytic Hydrogen on the Generation of Basic Chemicals, Chem.Eng.Progr., 63 (4), 59 (1967)
(24) M.G. Bowman and E.I. Onstott: Hydrogen Production by Low Voltage Electrolysis in a Combined Thermochemical and Electrochemical Cycle, Proc. Electrochemical Society Meeting, New York, 1974, Extended Abstracts, p 574
(25) G.H. Farbman: Studies of the Use of Heat from High-Temperature Nuclear Sources for Hydrogen Production Process, NASA CR-134918, NASA Lewis Research Center, January 1976

(26) L.E. Brecher, S. Spewock and C.J. Warde: The Westinghouse Sulfur Cycle for the Thermochemical Decomposition of Water, Int.J.Hydr.Energy, 2, 7 (1977)
(27) P.W.T. Lu and R.L. Ammon: Development Status of Electrolysis Technology for the Sulfur Cycle Hydrogen Production Process, Proc. 3rd WHEC, Tokyo, Japan, 23-26 June 1980, p 439
(28) C.J. Warde and L.E. Brecher: High Efficiency Cell for Hydrogen Production, Proc. 1st WHEC, Miami Beach, Florida, USA, 3-6 March 1976
(29) B.D. Struck, R. Junginger, D. Bolstersdorf and H. Neumeister: A Three-Compartment Electrolytic Cell for the Anodic Oxidation of Sulfur Dioxide and Cathodic Production of Hydrogen, Proc. 3rd WHEC, Tokyo, Japan, 23-26 June 1980, Abstract No 601
(30) K.E. Cox, W.M. Jones and C.L. Peterson: The LASL Bismuth Sulfate Thermochemical Hydrogen Cycle, Proc. 3rd WHEC, Tokyo, Japan 23-26 June 1980, p 345
(31) G.H. Schütz: Hydrogen Producing Cycles using Electricity and Heat-Hydrogen Halide Cycles: Electrolysis of HBr, Int.J.Hydr.Energy, 1, 379 (1976)
(32) G.H. Schütz and P.J. Fiebelmann: Electrolysis of Hydrobromic Acid, Int.J.Hydr.Energy, 5, 305 (1980)
(33) P.J. Fiebelmann and G.H. Schütz: Technological Aspects of HBr Electrolysis Development, Proc. 3rd WHEC, Tokyo, Japan, 23-26 June 1980, Abstract No 219
(34) G.H. Schütz and D.R. Lalonde: Proc. International Conference on Alternative Energy Sources, Miami Beach, Florida, USA, 5-7 December 1977
(35) G.H. Schütz: unpublished results
(36) G.E. Besenbruch et al: Hydrogen Production by the GA Sulfur-Iodine Process, A Progress Report, Proc. 3rd WHEC, Tokyo, Japan, 23-26 June 1980, p 243
(37) M. Dokiya, T. Kameyama and K. Fukuda: Thermochemical Hydrogen Preparation - Part V: A Feasibility Study of the Sulfur-Iodine Cycle, Int.J. Hydr.Energy, 4, 267 (1979)
(38) M. Dokiya et al: Thermochemical Hydrogen Produc-

tion by the SO_2-CH_3OH-I_2 Cycle, Proc. 3rd WHEC, Tokyo, Japan, 23-26 June 1980, p 373

(39) D.R. O'Keefe and J.H. Norman: The Liquid Hydrogen Iodide Decomposition Process Step for Water-splitting Applications, Proc. 3rd WHEC, Tokyo, Japan, 23-26 June 1980, p 277

(40) S. Sato, S. Shimizu, H. Nakajima and Y. Ikezoe: Nickel-Iodine-Sulfur Process for Hydrogen Production, Proc. 3rd WHEC, Tokyo, Japan, 23-26 June 1980, p 389

(41) H. Ishikawa, E. Ishii, I. Uehara and M. Nakane: Catalysed Thermal Decomposition of H_2SO_4 and Production of HBr by the Reaction of SO_2 with Br_2 and H_2O, Proc. 3rd WHEC, Tokyo, Japan, 23-26 June 1980, p 297

(42) K.F. Knoche, P. Schuster and T. Ritterbex: Thermochemical Hydrogen Production by a Vanadium/Chlorine Process, Proc. 3rd WHEC, Tokyo, Japan, 23-26 June 1980, p 551

(43) J.E. Funk and R.M. Reinstrom: System Study of Hydrogen Generation by Thermal Energy, Allison Division of General Motors, EDR 3714, Vol 11, Supplement A (1964)

(44) O.H. Krikorian: The ZnSe Thermochemical Cycle for Hydrogen Production: Chemical and Process Design Studies, Proc. 2nd WHEC, Zürich, 21-24 August 1978 p 791

(45) O.H. Krikorian and P.K. Pamela: The Utilization of $ZnSO_4$ Decomposition in Thermochemical Cycles, Proc. 3rd WHEC, Tokyo, Japan, 23-26 June 1980, Abstract No 226

(46) F. Behr, F. Flocke, K. Hunsänger, B. Kügler and W. Weirich: Alternative Hybrid Family Cycles, IEA Meeting, Tsukuba Science City, Japan, 18-21 June 1980

(47) F.J. Behr, F. Flocke, R. Schulten, H. Süssmann and W. Weirich: Development of the Li-H Cycle, Proc. 3rd WHEC, Tokyo, Japan, 23-26 June 1980, p 489

(48) A. Broggi and G. Beghi: Decomposition of Sulphuric Acid using Hot Air, 15th Intersociety Energy Conversion Eng. Conf., Seattle, USA, 17-22 August 1980

Conversion of solar energy into hydrogen and other fuels

by J. Gretz[1]

1. Introduction

Either free energy or a high temperature is needed to split water. Free energy can be supplied as light quanta (photolysis) or as electricity. The latter requires the conversion of solar radiation into electricity by thermomechanical or photoelectric means. Direct thermal decomposition (thermolysis) requires temperatures of several thousand degrees or - if thermochemical cycles are employed - temperatures of 600-800°C. In either case, the solar radiation has to be concentrated.

Photolysis by light quanta appears the most elegant solution but is unfortunately put in doubt by a number of intrinsic drawbacks. Not only has water a very low absorption coefficient for visible (ie energetic) solar radiation, but also biosystems alone (ie plants and algae) have the ability to split water. This they do using their complex chlorophyll system, comprising absorber (antenna), charge separator (membrane) and energy guide. Plants split water and manufacture chemical fuels such as cellulose from the resulting components together with atmospheric CO_2. But they do this with only a very low efficiency of eg 0.25-1%.

Whilst quantum conversion still requires much more fundamental research, solar electrolysis is already technically feasible. The question that remains is one of economics: what it now needs is technological development, and the construction of demonstration

1. Dr J. Gretz, Commission of the European Communities, Ispra Joint Research Centre, Ispra/Varese, Italy

power plants to set the iterative industrial learning process in motion and thereby make expensive solar energy competitive.

2. Solar radiation (1,2,3,4)

The Sun radiates energy to the value of

$$E_s = 3.8 \cdot 10^{23} \text{ kW}$$

with an energy density of

$$S_s = \frac{E_s}{\pi 4R^2} = 62600 \text{ kW/m}^2,$$

where R = radius of Sun = $6.95 \cdot 10^8$ m.

This energy density corresponds to a black-body temperature of

$$T_s = \sqrt[4]{S_s/\sigma} = 5760 \text{ K},$$

at the surface of the Sun, where σ = Boltzmann's constant = $5.67 \cdot 10^{-8}$ Wm^{-2} K^{-4}. The "quality" of this radiation is not reduced by the journey from the Sun to the Earth - ie with suitable equipment on the Earth's surface one can attain temperatures of almost 5760 K - but the radiation is "diluted". The problem of making efficient and economic use of solar energy lies precisely in this dilution, namely the low energy density of solar radiation at the surface of the Earth of approximately 1 kW/m^2 at most. This low figure is due to the small solid angle that the Sun appears to subtend from the Earth ($7 \cdot 10^{-5}$ steradians). The infra-red back-radiation, on the other hand, is radiated from the surface of the Earth over a solid angle of 2π (see Figure 1).

The energy harvest depends on geographic latitude and topographic altitude and is typically 2000 kWh m^{-2} $year^{-1}$ for Jerusalem, Texas and the Pyrenees and 800 kWh m^{-2} $year^{-1}$ for London.

The energy contained in light quanta participating in quantum reactions is expressed in mole quanta because according to Einstein's fundamental laws of photochemistry one molecule reacts for each photon absorbed.

The energy absorbed per mole is

$$\epsilon = N h \frac{c}{\lambda},$$

where N = Avogadro's number = $6.024 \cdot 10^{23}$ molecules mole^{-1}
h = Planck's constant = $6.63 \cdot 10^{-34}$ Joule sec
c = velocity of light = $3 \cdot 10^{8}$ m sec^{-1}
λ = wavelength of light, m.

Light wavelength of 680 nm is an absorption maximum for chlorophyll. At this wavelength the energy of one mole quantum is

$$\epsilon = 6.024 \cdot 10^{23} \cdot 6.63 \cdot 10^{-34} \cdot 3 \cdot 10^{8}/6.8 \cdot 10^{-7}$$

$$\epsilon = 176 \text{ kJ (42 kcal)}.$$

3. Quantum conversion (2,3,4,5)

Plants convert water and carbon dioxide into cellulose according to the global formula

$$CO_2 + H_2O + \text{light energy} \longrightarrow CH_2O + O_2.$$

In this conversion, water molecules are split by the photons of light, CO_2 is consumed, cellulose is produced and O_2 is released. Theoretically the reaction requires 470 kJ (112 kcal) per mole of O_2, ie

$$n = \frac{470}{176} = 2.7 \text{ mole quanta,}$$

at light wavelength of 680 nm.

Laboratory measurements using monochromatic light reveal a mole-quantum consumption of 8. This gives a theoretical efficiency of 8/2.7 ≃ 30% for photosynthetic conversion. However, in field conditions in the presence of polychromatic sunlight the actual efficiency is less by one or two orders of magnitude. This will be discussed in Section 4 "Bioconversion".

Photosynthesis consists of two main reactions: the light reaction, in which water is split

$$2\,H_2O \rightarrow 4\,H^+ + O_2 + 4\,e,$$

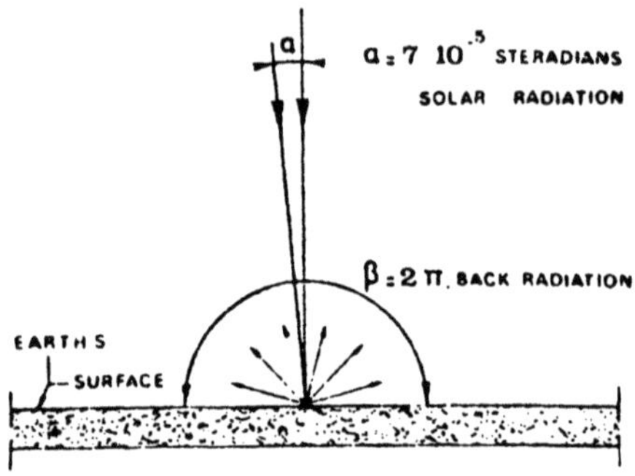

<u>Figure 1</u>
Incident solar radiation and back-radiation

and the dark reaction, in which the hydrogen ions (protons) are attached to the CO_2 to form cellulose (see Figure 2)

$$H_2{}^{+} + CO_2 \rightarrow CO_2O + O.$$

The actual energy required during the light reaction to split one mole of water in liquid form is 237 kJ (56.7 kcal). A photon of wavelength

$$\lambda = \frac{280\,000}{56.7} = 495 \text{ nm}$$

(Wien's displacement law) is therefore capable of splitting water. As stated above this does not happen. Rather, the energy of 8 mole quanta (1400 kJ) needs to be collected and used, since energy is lost in the chlorophyll machinery.

Nevertheless it might be possible - and this is the target most widely adopted - to prevent the H^+ proton from attaching to the CO_2 and to deflect it away where it can form hydrogen. However, the times taken for the separate degradation steps are extremely short and make intervention difficult. For example, in vivo the average lifetime of chlorophyll in its excited state is $1.6 \cdot 10^{-9}$ sec. Thus, plants use their complex enzymatic chlorophyll system (consisting of absorber, charge separator (membrane) and energy collector) to collect and process several quanta of light, even though with only extremely low efficiency. In

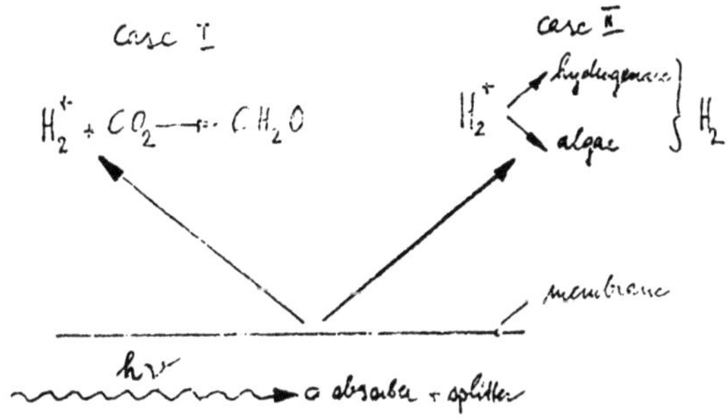

<u>Figure 2</u> Quantum conversion

1973 hydrogen was produced using an in vitro system comprising isolated chloroplasts and the enzyme hydrogenase that was irradiated with sunlight (see Figure 2). In 1974 hydrogen was produced using blue algae. However the yield and lifetime of these systems were low.

Hydrogen produced by direct quantum conversion might be achieved by making artificial photo-systems using synthetic catalysts and membranes that separate the unwanted dark reaction from the light reaction and allow the latter to produce hydrogen directly.

4. Bioconversion (4,5,6,7)

Bioconversion is the use of case I in Figure 2, ie plant carbohydrate production and its utilization in various processes, eg pyrolysis, fermentation, and gasification. As mentioned in Section 3, the efficiencies are poor but the product is valuable since it represents stored energy.

I shall not discuss here the various technologies by which cellulose is converted into chemical products, because they have been more or less open secrets within the chemical industry for some time. I shall instead confine myself to one concrete example to calculate the efficiency of bioconversion.

A typical, mixed forest in central Europe produces 1.25 kg of wood* per square metre surface area per year. By anaerobic fermentation, 0.5 m^3 of fuel gas* (50-70% methane) can be produced from 1 kg of wood. This fuel gas has a calorific value of 5300 kcal/m^3*. The efficiency of this forest is therefore

$$\eta = \frac{E_{out}}{E_{in}} = \frac{5300 \cdot 0.5 \cdot 1.25}{1500 \cdot 860} = 0.26\%$$

assuming average solar radiation of 1500 kWh m^{-2} $year^{-1}$.

Sugar-beet, maize, soya beans and a number of other plants have higher efficiencies of up to 3% but even these are low. Bioconversion is not efficient.

The use of organic wastes is sometimes classed under bioconversion since fresh organic materials and organic wastes ferment in practically the same manner. The arithmetic gives a curious coincidence: each citizen of Europe produces 1.5 kg of waste per day. Half of this is organic in nature. Each citizen therefore produces from his waste

$$E_{waste} = \frac{0.75 \cdot 2600}{860 \cdot 24} \simeq 100 \text{ W}$$

of energy - which is equal to his own quiescent metabolic energy.

5. Electrolysis (8,9,10,11,17)

Electrolysis is a more attractive way of producing hydrogen with solar radiation since it can be operated intermittently and therefore needs no storage.

The solar electricity needed for electrolysis can be produced either photoelectrically or thermomechanically. Both technologies are available today. Solar electrolytic hydrogen production is therefore a question of the cost of components and of development.

---

*average values

Type of installation and boundary conditions

Experimental, with central radiation receiver
Site: 37.6°N, 14.8°E (Adrano/Catania/Sicily)
Alignment point: same day and night, midday, radiation accepted 100 W/m²

Main data

Nominal power	1 MW (el)
Thermal power at receiver	4.8 MW (thermal)
Reflecting area, total	6216 m²

Heliostat array

Heliostats:	two types steerable about two axes precision ± 4 mrad 1σ
CETHEL type:	approx. 52 m², 8 focusing modules, 70 heliostats
MBB type:	approx. 23 m², 16 focusing squares, 112 heliostats

Receiver/tower

Cavity receiver: aperture 4.5 m Ø at a height of 55 m, inclination 110°
Steam at output of receiver: 512°C, 64 atm, 4860 kg/h

Steam circuit

Turbine with receiver connected, no heat-exchanger between
Rated power: 1200 kW (mech) with steam at 510°C, 60 atm
Feed water temperature at receiver input: 36°C
Cooling water temperature: 25°C maximum

Thermal storage

For running for 30 minutes at reduced power
Storage media: water/steam 300 kWh; salt (Hitec) 60 kWh
Pressurized water tank (19 bar) for 4300 kg, steam at 19 bar to 7 bar
Salt storage: two tanks for a total of 1600 kg Hitec
Heat-exchanger for 19 bar, 480°C and 410°C steam

Electrical system

3-phase generator for 1100 kW minimum
Externally available: 100 kW; consumption within plant: 100 kW
Transformers/emergency power supply for pumps, connection to mains: switch, transformer, etc
Controlled from central control room: steam circuit, contollers for power-station subsystems

Table 1 Technical parameters of the EURELIOS solar thermal power plant

The EURELIOS solar thermal power plant is an example of a solar electricity generating station. This power plant was constructed for the EEC by a consortium of European companies (MBB, Federal Republic of Germany; ANSALDO, Italy; CETHEL, France) and completed in December 1980. It is therefore the first large solar power plant in the world and will supply current this year (1981) into a main distribution grid (ENEL, Italy). The station produces 1 megawatt, which is sufficient to supply a small village with electricity.

EURELIOS is a demonstration solar thermal power plant and uses a reflector area of 6200 m^2 to concentrate solar radiation on to a boiler mounted on a 55 m high tower. The steam produced in the boiler drives a turbogenerator that supplies the electricity.

Table 1 gives the plant's technical parameters, and Figure 3 shows a photograph of it.

Figure 3 EURELIOS

When the breakthrough in high-temperature electrolysis comes it could lead to an interesting application for thermomechanical solar electricity generation: only part of the coolant heated in the receiver is released into the turbogenerator; the remainder is used to supply the $T\Delta S$ (thermal component) of the electrolyser (see Figure 4). The cell voltage increases from $E = 1.49$ V at 25°C to $E' = 0.88$ V at 1000°C for the same current density in the electrodes, owing to the voltage/temperature gradient of

$$\frac{dE}{dT} = 0.25 \text{ mV, } {}^\circ C^{-1}.$$

The difference $\Delta E = 1.49 - 0.88 = 0.58$ V can be supplied by enthalpy ($T\Delta S$), ie by energy not subject to Carnot-cycle degradation.

Savings can be made with such hybrid electrolysis systems (see Figure 5). The following calculation shows that those savings can be considerable (theoretical values, with mechanical efficiency $\eta_m = 1$, overvoltage = 0 V). From Figure 5 the heat flowing Q_g to the turbogenerator is

$$Q_g = [\frac{88}{\eta}] (\frac{88}{\eta} + 58)^{-1}] \cdot 100 = 88\,\%,$$

assuming an overall conversion efficiency of heat into electricity of 20%. Under these conditions the savings in the cost of producing hydrogen expressed in mills*/kWh would be, compared with conventional current electrolysis:

$$\Delta\omega = \frac{a \cdot 10^3}{S_y} [F (\frac{i_1}{\eta} - i_1 + i_2)],$$

where

$\Delta\omega$ = difference in costs of producing hydrogen by hybrid electrolysis and conventional electrolysis
α = (0.16) = annuity, year^{-1}
F = (0.12) = enthalpy component
S_y = (2300) = annual hours of sunshine, h year^{-1}
i_1 = (200) = capital cost of thermomechanical solar power station, $ kW(th)$^{-1}$
i_2 = (120) = capital cost of turbogenerator, $ kW(el)$^{-1}$
i_3 = (120) = capital cost of electrolyser, $ kW(el)$^{-1}$
η = (0.2) = efficiency of conversion into electricity.

The numbers in brackets are values assumed for 1985-1990 and expressed in 1980 monetary terms.

*$ 10^{-3}

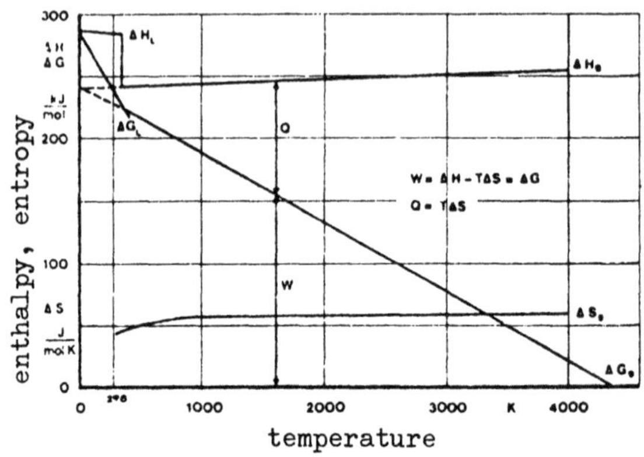

<u>Figure 4</u> The thermodynamics of electrolysis

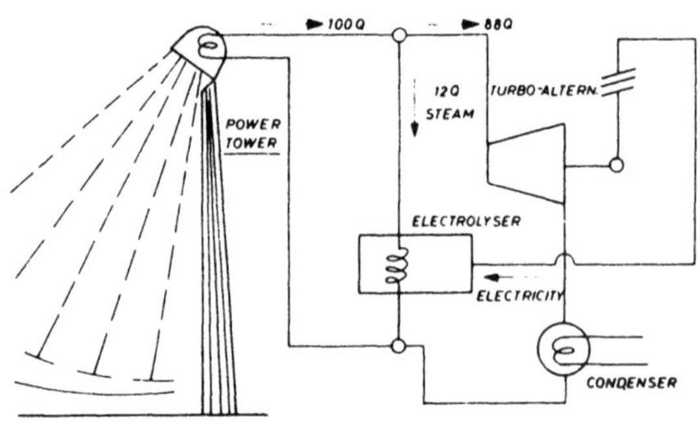

<u>Figure 5</u> Hybrid electrolysis

The cost of producing hydrogen by conventional electrolysis is:

$$\omega = \frac{a \cdot 10^3}{S_y}\left[\frac{i_1}{\eta} + i_2 + i_3\right] = \frac{160}{2300}(1000 + 120 + 120) = 87 \text{ mills/kWh.}$$

The difference in cost between this method and hybrid electrolysis gives a theoretical saving of

$$\Delta\omega = \frac{160}{2300}\left[0.12\,(1000 - 200 + 120)\right] = 7.7 \text{ mills/kWh},$$

ie approximately 8.8%.

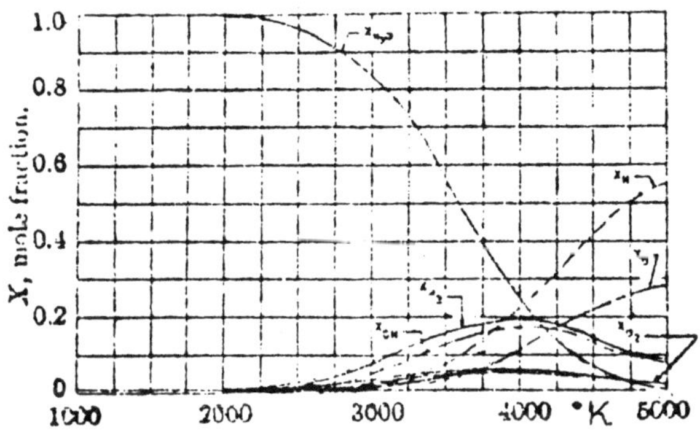

Figure 6 Composition of dissociated water, p = 10 atm

A critical cost/benefit analysis would have to be undertaken to show whether the gain in energy costs in the hydrogen production process offsets the increased investment required for high-temperature electrolysis. A reminder must be given here that all terrestrial solar energy systems with annual load factors of at best 25% have poor amortization returns, which favours low capital costs rather than high efficiencies.

6. Thermolysis (9,12,13,14,15)

At elevated temperatures water splits into the following components via the reactions shown (see Figure 6):

$$H_2O \rightarrow \frac{1}{2}H_2 + OH$$

$$H_2O \rightarrow \frac{1}{2}O_2 + H_2$$

$$\frac{1}{2}H_2 \rightarrow H$$

$$\frac{1}{2}O_2 \rightarrow O.$$

Whilst temperatures up to ~4500 K can be attained in solar installations, the difficulties experienced with constructional materials at these temperatures and the separation of hydrogen from such a hot mixture makes the use of direct thermolysis still impossible today.

Thermolytic processes ought to be realizable using pure solar technology. For example, at T = 3000 K the mixture consists of 84% water (see Figure 6). Since H_2 is diathermal, the greater part of this mixture reacts with the radiation energy predominant at 3000 K. The heat transfer relationships of H.C. Hottel and R.S. Egbert for $0.002 < pxd < 0.7$ can be applied to the conditions prevailing in the receiver, where p is the partial pressure of water and d is the layer thickness. Applying these relationships gives radiative heat-transfer coefficients of the order of 500×10^3 kcal m^{-2} h. With such a heat transfer coefficient it should, for example, be technologically possible to incorporate the "piping" inside a cavity receiver of 16 m^2 opening (5 MW thermal power).

7. Thermochemical decomposition (9,13,16)

The attainment of temperatures of up to 4500 K obviously simplifies the search for suitable thermochemical circuits. Such circuits are currently confined to the use of gas reactors, ie for temperatures of $650^\circ C < T < 950^\circ C$.

However, the intermittent nature of solar radiation implies the need, in contrast to electrolysis, either for thermal storage or for additional heating for the periods when radiation is absent. This raises a fundamental consideration.

Any terrestrial solar energy installation labours under the handicap of low amortization returns on account of its inherently poor load factor of 25% at best (average 15%). Storing the energy is no solution since the installation is designed in proportion to performance and not in proportion to energy, ie the total performance-proportional part of the solar power station (heliostats, tower, receiver, primary circuit - ie approximately 50-60% of the total installation) is only amortized to 15-25%, with or without storage. In addition, the storage system will require investment and a return will be wanted on it too. This investment is greater than that for a hybrid oil burner. A hybrid oil burner running in parallel would therefore be more economical than a storage system. In view of this, the likely trend will be to run the solar conversion system in parallel with a fossil-fired installation as a "fuel saver", at least during a possible preliminary period of transition from the era of fossil fuels to that of alternative energies, and not to operate the solar hydrogen production plant autonomously with solar energy.

Nevertheless the high temperatures attainable in concentrating solar power stations offer advantages as regards the chemical reactions they can promote. For example, gaseous SO_3 can be split solar-endogenously into SO_2 and O_2 and the two components then transported separately to a place of use and there recombined releasing energy in the process. Not only does the proportion that dissociates increase with increasing temperature but the pressure (at eg 950 K) at which the process occurs is considerably lower than that inside a gas reactor at its working temperature.

8. Outlook for the future (16)

Unfortunately, things seem somewhat pessimistic if an attempt is made to predict when solar hydrogen or solar energy in general will contribute any substantial proportion of our primary energy requirement.

Figure 7, prepared by IIASA, shows the market penetration of various primary energy sources in the USA. The slopes of all the curves are roughly equal, ie 50% of the market in 100 years, irrespective of infrastructure and changes in technology and society. Solar energy - and with it presumably an amount of solar hydrogen - is expected to meet 50% of the demand for primary energy, along with nuclear fusion, by the year 2080 if one regards 1980 as the year when solar energy and fusion began to be used. This is also roughly the prediction of the Commission of the European Communities (see Figure 8). But at least two things must be said in this context.

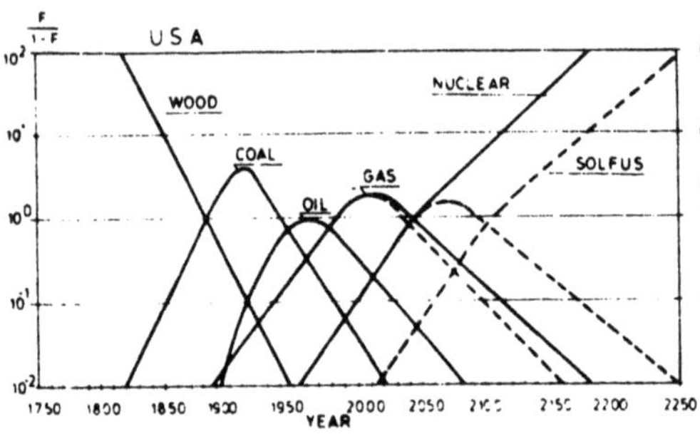

Figure 7 Primary energy sources in the USA

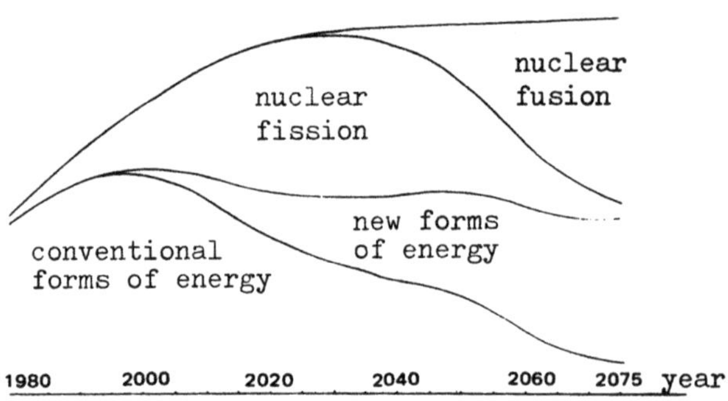

Figure 8 EEC forecast for primary energy sources

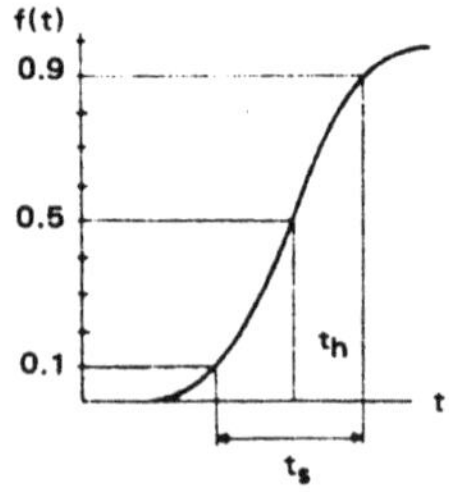

Figure 9 Logistic penetration function for primary energy sources

Firstly, the logistic market penetration function is applicable only over the time t_s (see Figure 9) during which the market share grows from 10% to 90%. It takes 20-30 years before the 10% limit is reached, which means that the time when solar energy and fusion will attain 50% of the market would be shifted to between 2100 and 2110.

Secondly, the primary energy source at any given time has replaced its predecessor on the grounds of lower price (or cost) and more convenient use - neither of which is the case for solar energy or, in extenso, for solar hydrogen. Solar energy has many advantages but is certainly not cheaper and more convenient than oil or nuclear energy.

All in all, the author is of the view that when, if ever, solar energy comes into use on a large scale, some of this will be in the form of fuels produced by solar means, and hydrogen will hold a key position amongst these fuels. Whether this happens or not depends less on the technology of these decentralized primary energy sources and more on socio-political imponderables that cannot be quantified.

9. Summary

Hydrogen obtained by using solar energy to split water is an inexhaustible source of transportable primary energy. Solar hydrogen is also an ideal storage medium for solar energy.

This paper presents a number of energy characteristics of solar radiation and describes the thermodynamic and energetic conditions of splitting water. The various ways of splitting water are then considered, namely quantum conversion, bioconversion, electrolysis, and thermal conversion.

An attempt is made to predict the extent to which solar hydrogen will contribute to meeting our energy demand.

References

(1) J. Gretz: The Conversion of Solar Energy without Concentration, Energie Nucleare, 21 (8/9) (1974)
(2) D.O. Hall: Solar Energy Use through Biology - Past, Present and Future, Solar Energy, 22 (307-328) (1979)
(3) E. Broda: Photochemical Production of Hydrogen from Water, Second World Hydrogen Energy Conference, Zürich (1978)
(4) D.O. Hall: Solar Energy Conversion through Biology - Could it be a Practical Energy Source? Fuel, 57, June (1978)
(5) Sir G. Porter and M.D. Archer: In Vitro Photosynthesis, Interdisciplinary Science Rev., 1 (2) (1976)
(6) J. Bockris: Energy: The Solar Hydrogen Alternative, Architectural Press, London (1975)
(7) J. Gretz: Bioconversion of Solar Energy, Internal Working Paper, Ispra (1974)
(8) Die Nutzung der solaren Strahlungsenergie (Utilization of solar radiative energy), published by the Federal Ministry of Research and Technology, Federal Republic of Germany (1976)
(9) J. Gretz: On the Potential of Solar Energy Conversion into Hydrogen and/or other Fuels, Int.J. Hydrogen Energy, 5 (269-280) (1980)
(10) A Hydro-Energy Carrier, 2, NASA-ASEE (1973)
(11) Hydrogen Energy, Part A, University of Miami (T. Nejet Veziroglu, Ed), Plenum Press, New York (1975)
(12) B.J. East Lund and W.C. Gough: Generation of Hydrogen Ultraviolet Light Produced by the Fusion Torch, 16th National Meeting Am.Chem.Soc., Boston Massachusetts (April 1972)
(13) J.E. Funk and R. Reinstrom: Energy Requirements in the Production of Hydrogen from Water, I & EC Process Design and Development, 5 (3), July (1966)
(14) P.J. Friel and R.C. Goetz: The Composition and Enthalpy of Dissociated Water Vapour, J.Phys. Chem., January (1960)
(15) A. Schack: Der industrielle Wärmeübergang (Industrial heat transfer), Verlag Stahleisen, Düsseldorf

(16) W. Häfele et al: IIASA, Second Status Report of the IIASA Project on Energy Systems, (1975)
(17) EURELIOS, The 1 MW Experimental Solar Thermal Power Plant of the European Economic Community, EUR 6247 (1980)

Cost comparisons of different methods of producing hydrogen

by D. van Velzen[1]

1. Introduction

The so-called hydrogen energy system is closely linked to the development of new energy sources. However, we should like to stress here that hydrogen itself cannot in any way be regarded as a substitute energy source. The hydrogen energy concept rests on known alternative energy sources such as coal, nuclear energy and (perhaps in the future) solar energy as well. In order to integrate these energy sources into the overall energy consumption system, a new energy vector is needed. Under certain circumstances this new energy vector could be hydrogen. The hydrogen energy concept is therefore nothing other than a future energy distribution system in which hydrogen is used on a large scale as a heating gas in industry and the home and as a fuel in the transport sector.

It is logical to ask whether the hydrogen energy system can ever be realized and if so, when. This question can be answered only by comparing the future availability of present-day energy vectors (petroleum and natural gas) with the situation for hydrogen. Estimates of the cost of future processes for producing hydrogen are required for this comparison. A large number of such estimates have already been made but their results are not always unambiguous. There are large discrepancies because of the different hypotheses on which most studies are based.

---

1. Chem.Ing. Micheme D. van Velzen, Commission of the European Communities, Ispra Joint Research Centre, Ispra/Varese, Italy

In general, costs are analysed in the following way:

- Different hydrogen production processes are compared on the basis of different feedstocks, eg petroleum, natural gas and coal. The current overall costs for the processes are determined, including the proportion represented by the cost of the primary energy.
- A model describing movements in energy prices is constructed to allow extrapolation into the future. Assumptions about expected improvements in new technologies are also taken into account in this model.
- Future hydrogen production costs are then calculated using the above energy model.
- These calculations yield a number of graphs showing the costs of producing hydrogen by different processes as a function of time. The idea is that one can then read from these graphs which technology has the best prospects and when it will become competitive.

These conclusions are governed largely by the type of model. It therefore follows that different models can lead to different results. Consequently, such studies cannot be very precise and their findings must always be treated with the greatest care.

2. Method

No energy model is employed in this study. Only the influence of rising energy prices on the cost of producing hydrogen is considered. We hope that this type of analysis will contribute to a better understanding of costing.

We examined the costs of hydrogen production using plant with a standard production capacity of 100 000 Nm^3/h (energy production: approximately 10 PJ/year). In this we assume that we can divide the costs of hydrogen production into only two components:

a. the cost of the consumption of primary energy (coal, petroleum, etc): we call this factor energy cost component;

b. all other costs, such as capital costs, handling, operating costs, etc, are grouped under the heading operating cost component.

The operating cost component is kept constant on the basis of 1979 prices, since no reliable method exists for taking the effect of rising energy prices into account in these costs. The energy cost component is increased by as much as five times from the baseline value.

We are well aware that fixing the operating cost component on the basis of 1979 prices leads to minor errors, in that the effect of rising energy prices on operating costs is ignored. However, the general direction of price movements is not essentially affected by adopting this method.

3. Processes for producing hydrogen

At present, hydrogen is used in large quantities both in the fertilizer industry and in the petrochemical industry. Total consumption within the EEC is roughly 4 million tonnes per year, of which 60% is used for synthesizing ammonia and 26% by the oil industry for hydrocracking and sulphur removal. There are various ways of producing hydrogen industrially. The steam reforming of natural gas and the partial oxidation of heavier petroleum fractions are the processes used most frequently today. All present-day processes for producing hydrogen consume fossil fuels.

Increasing demand, together with the possible use of hydrogen as an energy vector, have led to intensified research into new technologies for producing hydrogen. The new processes no longer use the fossil fuels petroleum and natural gas. Instead, they use alternative energy sources such as coal, nuclear power, solar energy and biomass. Numerous processes have been proposed and are currently at various stages of development.

As an example, we shall take the production of hydrogen from coal. Four different processes are used at present in the field of coal gasification and a further 20 or so are being studied in industrial and semi-industrial experimental plant. Similarly there are numerous new developments in the field of water electrolysis. In view of the large number of new processes we find ourselves forced to restrict this study to six representative processes. These processes were chosen on the basis of the energy sources used:

- petroleum,
- natural gas,
- coal,
- electricity,
- nuclear energy,
- solar energy.

For each source of energy we shall use only one typical process. We can make this simplification because there are in general only relatively small price differences between the individual processes based on any one energy source and because predicting the cost of new technologies is always imprecise. Table 1 summarizes the results of the study.

1. Hydrogen from petroleum

The cost of producing hydrogen by processes in this category is calculated from data obtained from existing large industrial plants that have been operating for many years. The partial oxidation of heavier petroleum fractions serves as an example. In this process oil, steam and oxygen are introduced together into a reactor and partially oxidized in a flame at a relatively high temperature. The existing molecular structure is completely destroyed and a gas having the following composition is produced:

CO_2 = 4.9 vol.%
CO = 46.2 " "
H_2 = 46.0 " "
CH_4 = 0.6 " "

and 2.3% of other products (H_2S, N_2, COS).

Table 1 Estimated hydrogen production cost
(all figures are expressed in (1979) $/GJ)

Process	Partial oxidation	Steam reforming	Coal gasification	Electrolysis		Thermochemical (hybrid) cycles
Primary energy source	heavy oil	natural gas	coal	nuclear energy		nuclear energy
Primary energy price (1979)	2.9	3.0	1.5	0.42		0.42
				Today	Future	
Efficiency (%)	76	72	60	27	31	40
HYDROGEN PRODUCTION COST						
Operating cost component	1.9	0.9	3.5	9.1	6.5	7.3
Production cost (1979)	5.7	5.1	6.0	10.6	7.8	9.4
idem with energy prices doubling	9.5	9.2	8.5	12.2	9.2	9.4
idem with energy prices tripling	13.3	13.4	11.0	13.7	10.6	10.5
idem with energy prices increasing 5-fold	20.9	21.7	16.0	16.8	13.3	12.6

The 46 vol.% of CO is then converted into hydrogen in a second stage at 450°C:

$$CO + H_2O \rightarrow CO_2 + H_2 .$$

Carbon dioxide is separated out and the remaining CO is converted into methane and water in the presence of catalysts.

The production cost for this process is very closely dependent on the price of the raw materials, as can be seen in Figure 1. This figure shows the results of a study conducted at the request of the International Energy Agency (1). In this study the future cost of producing hydrogen using the same process was estimated by experts from different countries. It can be seen that the cost estimates vary widely even for this very well-known process. The standard deviation of the data in Figure 1 is 8.7%. Today's price (1979) for the heavy oil used in the process is $ 2.9/GJ and the efficiency of the process is 76%. These figures give the following estimate for the cost of producing hydrogen:

energy cost component (2.9/0.76)	$ 3.8/GJ
operating cost component	$ 1.9/GJ
total cost	$ 5.7/GJ

If the primary energy price doubles from $ 2.9/GJ to $ 5.8/GJ, the cost of producing the hydrogen increases to $ 9.5/GJ. A tripling of the price brings the production cost to $ 13.3/GJ.

5. Hydrogen from natural gas

The steam reforming of natural gas is at present the most economic method of producing hydrogen. The process is based on the known reaction between steam and methane (or other hydrocarbons):

$$CH_4 + H_2O \rightarrow CO + 3\,H_2 .$$

The reaction takes place at a temperature of approxi-

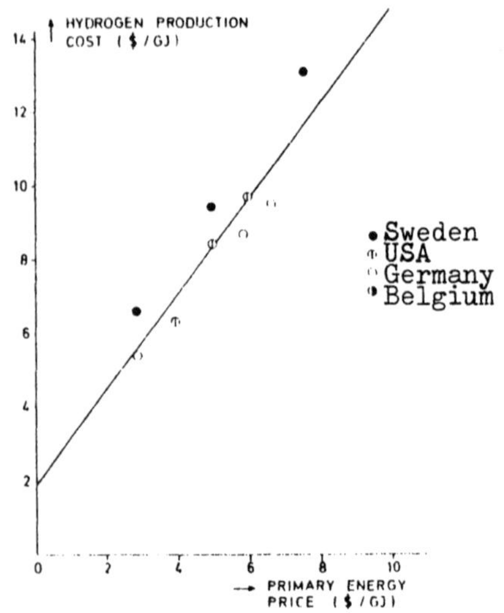

Figure 1 Hydrogen production cost
Partial oxidation of heavy oil

mately 800°C and a pressure of approximately 20 bar. As with partial oxidation, the CO component is again converted with steam into CO_2 and hydrogen.

In this case too, plenty of data on hydrogen production costs are available from existing plant using this large-scale industrial process. Figure 2 shows the results from the study (1) mentioned earlier. The figure shows the hydrogen production cost as a function of the price of natural gas. With this process, the operating cost component for the plant is lower than that for the partial oxidation of heavy oil:

steam reforming	$ 0.9/GJ
partial oxidation	$ 1.9/GJ.

The average price (1979) of natural gas is $ 3.0/GJ and the efficiency of steam reforming is approx. 72%. These figures give the following cost breakdown for the process:

energy cost component (3.0/0.72)	$ 4.2/GJ
operating cost component	$ 0.9/GJ
total cost	$ 5.1/GJ

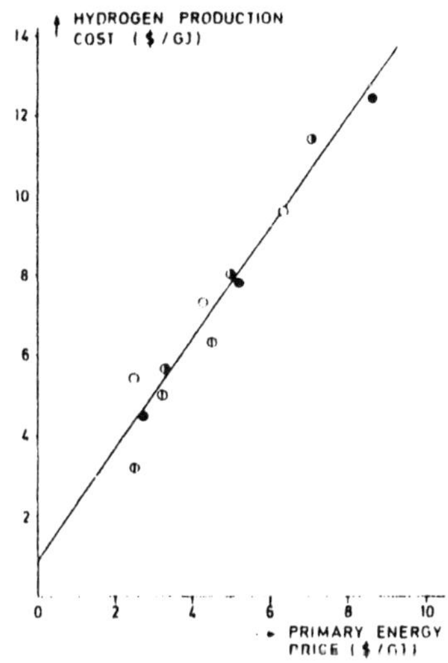

Figure 2 Hydrogen production cost
Steam reforming of natural gas

If the price of natural gas rises from $ 3.0/GJ to $ 6.0/GJ, the price of hydrogen increases from $ 5.1/GJ to $ 9.2/GJ. If the price of natural gas triples, the price of hydrogen rises to $ 13.4/GJ.

Today these two processes, steam reforming and partial oxidation, produce more than 80% of the total amount of hydrogen needed. Both processes are similar in cost, the main cost component being the cost of the energy consumed. The energy cost for partial oxidation is 67% of the total cost, and for steam reforming it is 83%. The difference in cost between the two processes is small. Which process is to be preferred will depend on the location of the plant. The location is influenced by specific factors such as the presence of an infrastructure and the availability of either natural gas or heavy oil.

6. Hydrogen from coal

In principle the gasification of coal is partial oxidation, similar to the oxidation of heavy oil fractions. The disadvantage of the process is the need to move large quantities of solid material and the removal of a considerable amount of ash. Various industrial plants are currently being operated and numerous new processes are being prepared: these new processes are at various stages of development. The large number of processes in existence makes a precise analysis of costs very difficult. In addition it is assumed that coal gasification plants will be operating at their most efficient with a production capacity of 1 200 000 Nm^3/h of H_2 (energy production 120 PJ/year). This figure is 10% higher than is normally the case today.

These considerations make cost calculations on a standard basis almost impossible. Consequently the published cost estimates for future coal gasification plants are very different. For example, the IEA study mentioned earlier grouped together the cost estimates made by various countries. This revealed forecasts for 1985 varying widely between $ 3.6/GJ and $ 7.8/GJ.

In order to make use of these different estimates in our study we felt obliged to ignore the extreme figures and to work with an average derived from the more reliable figures. We also used a study by Ekman (2), examining the situation in the USA. Most of the data employed refer to the Koppers-Totzek and Lurgi processes. The following conclusions can be drawn:

- The efficiency of the coal gasification process is lower than that of steam reforming and of partial oxidation of heavy oil:

  55-60% for coal gasification,
  65-75% for reforming and partial oxidation.

We have used 60% in our calculations here.

- The total capital cost of a plant with a capacity of approx. 100 PJ/year is roughly $ 2000 million, ie $ 20/GJ/year.

- This capital cost leads to an operating cost component of approx. $ 3.5/GJ.

- The primary energy source (coal) is much cheaper than natural gas and heavy oil, being $ 1.5/GJ as against $ 3.0/GJ.

These figures give the following cost breakdown for the process:

energy cost component (1.5/0.6)	$ 2.5/GJ
operating cost component	$ 3.5/GJ
	$ 6.0/GJ

If the price of coal doubles from $ 1.5/GJ to $ 3.0/GJ the price of hydrogen becomes $ 8.4/GJ. If the price of coal triples, the hydrogen costs $ 10.9/GJ.

At present, steam reforming and partial oxidation are still cheaper hydrogen production processes than coal gasification. As energy prices continue to rise this situation will change in favour of coal gasification processes.

7. Water electrolysis

The electrolysis of water is a simpler and cleaner process: hydrogen and oxygen are separated and collected in their pure form. The great disadvantage of electrolysis is its use of very expensive energy - electricity. Furthermore, this electrical energy is not primary energy but must itself be generated by consuming primary energy, which adversely affects the efficiency of the process chain.

Electricity can be generated from almost all primary energy sources: heavy oil, natural gas and coal dominate the market at present.

However, it is obvious that a sequence of processes such as the conversion of heavy oil into electricity and electricity into hydrogen cannot be an economic method of producing hydrogen, since the direct route by steam reforming or partial oxidation will always have the advantage. The capital costs and the efficiency of electrolysis are significantly less favourable. Consequently, electrolysis can never be a favoured process for hydrogen production as long as the electrical energy is generated using fossil fuels.

For these reasons, in this section we shall consider only electrolysis using electricity generated by nuclear energy.

Our analysis is based mainly on the work of a group of European experts (3). This defined a general method and specified the range of process parameters that served as the basis for analysing the cost of producing hydrogen. They assumed the use of pressurized water reactors (PWR) and high temperature reactors (HTR) for generating the electricity. The costs of generating electricity with both types of reactor are almost identical ($ 5.3 vs. $ 5.6/GJ or 19 vs. 20 mills/kWh) even though the reactors are different and have different efficiencies (33% vs. 38%) and different capital costs ($ 650 vs. $ 765/kW_e). We therefore used an average of $ 5.5/GJ for calculating the cost of the electricity.

At the moment, water electrolysis is not an economically competitive process. A large number of possible improvements to the process are being investigated. This research is currently on a laboratory scale or at best being studied in small experimental plants.

The following improvements are being worked on for alkaline electrolysis:

- development of new electrocatalysts for the anode;
- improving diaphragms and membranes;
- using higher temperatures and pressures;

- increasing the current density.

Another new development is the solid-polymer electrolyte (SPE) system. Here the electrolyte consists of a solid semi-permeable Nafion membrane possessing the properties of a cation exchanger. When the membrane is saturated with water, it becomes an ion conductor with a low electrical resistance.

There are no reliable cost data for future methods of producing hydrogen electrolytically. Costs can be estimated either on the basis of existing small-capacity commercial plants or on the basis of optimistic extrapolations to future improved processes. Many of these optimistic estimates are based on targets for future process performances and costs that rarely rest on proven experimental data. The aforementioned working group (3) assumed a cell voltage of 1.6-1.7 V at a current density of 10-15 kA/m^2 for calculations regarding future processes. However, the latest results with SPE electrolysis using cells of 0.23 m^2 show cell voltages of 1.95-2.0 V at this current density (4). Consequently, in our cost analysis we have used both present-day data and optimistic performance data.

There are fundamental differences between the electrolysis of water and the production of hydrogen from fossil fuels (this applies to the present as well as to the future).

- The primary energy used - nuclear power - is much cheaper than energy from fossil fuels. Today's price (1979) for uranium fuel is approx. $0.42/GJ, as against $ 1.5/GJ for coal and $ 3.0/GJ for natural gas and petroleum.

- A large amount of capital has to be invested in a nuclear power station: in our case, roughly $ 4.3/GJ. The energy cost component is $ 1.2/GJ assuming a 35% efficiency.

  This means that for the generation of electricity by nuclear power the energy cost component is only 25%.

Today's commercial process for electrolysing water operates with a cell voltage of approx. 1.9 V and a power consumption of 4.65 kWh/m^3 of H_2. This is equivalent to an efficiency of 76% for converting electricity into hydrogen. If we assume that the efficiency for generating the electricity is 35%, the overall efficiency of the nuclear power/electricity/ hydrogen process chain is 27%.

The operating cost component for present-day electrolysers has been calculated at $ 3.4/GJ. These figures therefore give the following cost breakdown for hydrogen production:

energy cost component (0.42/0.27)		$ 1.6/GJ
operating cost component		
nuclear reactor (4.3/0.76)	5.7	
electrolyser	3.4	
		$ 9.1/GJ
total cost		$10.7/GJ

The following improvements are assumed for future electrolysis:

The cell voltage will drop to 1.65 V. This means an increase in electrolysis efficiency from 76% to 89% and in overall efficiency from 27% to 31%. This increase in efficiency will reduce the hydrogen production cost from $ 10.7/GJ to $ 9.6/GJ.

Another very optimistic assumption is a reduction in the capital cost of future electrolysis plant from $ 3.4/GJ to $ 1.65/GJ. If the two improvements are combined, the following costs breakdown is obtained:

energy cost component (0.42/0.31)		$ 1.3/GJ
operating cost component		
nuclear reactor (4.3/0.89)	4.8	
electrolyser	1.7	
		$ 6.5/GJ
total cost		$ 7.8/GJ

Thus, the energy cost component is then no more than 17% of the total cost. This is very low compared with the production of hydrogen from fossil fuels, where it is 50-80% of the total. The consequence is that a rise in the primary energy price has relatively little effect on the cost of producing hydrogen. If the primary energy price doubles from $ 0.42/GJ to $ 0.84/GJ, the hydrogen cost becomes $ 9.2/GJ; if the price triples, the hydrogen cost becomes $10.6/GJ.

8. Production of hydrogen from thermochemical and hybrid processes

The thermochemical processes for producing hydrogen are cyclic processes with two or more reactions that split water into hydrogen and oxygen with the application of heat:

$$H_2O + \text{energy} \longrightarrow H_2 + 1/2\ O_2.$$

The primary energy passes directly into these processes without conversion into electricity. In addition to closed, purely thermochemical processes, what are known as "hybrid" processes have been developed, combining chemical and electrochemical stages.

Today's most highly developed thermochemical processes are both hybrid processes:

- Westinghouse sulphur cycle (Ispra Mark 11):

  $SO_2 + H_2O \rightarrow H_2SO_4 + H_2$ (electrochemical)
  $H_2SO_4 \rightarrow H_2O + SO_2 + 1/2\ O_2$ (1070 K)

- Mark 13 sulphur/bromine cycle (Ispra Joint Research Centre):

  $2\ HBr \rightarrow H_2 + Br_2$ (electrochemical)
  $2\ H_2O + Br_2 + SO_2 \rightarrow 2\ HBr + H_2SO_4$ (360 K)
  $H_2SO_4 \rightarrow H_2O + SO_2 + 1/2\ O_2$ (1070 K).

Both processes have been tested experimentally in the laboratory. An experimental plant for 10 m^3/h of H_2 is being constructed. No reliable data for the production of hydrogen using thermochemical cyclic processes are therefore available. Costs can only be estimated from related chemical processes. Various methods are known and all are based on many years' experience in the chemical industry. This experience allows relatively precise estimates to be made for the capital cost of chemical reactors, heat exchangers, piping and instrumentation. The precondition is that a complete flow diagram of the process be available and the reaction conditions well known.

A high-temperature nuclear reactor (HTR) is envisaged as the primary energy source for the above thermochemical processes. Cost data for the HTR are as yet based on estimates and therefore uncertain. This is a very important point because the capital costs for the nuclear part are the largest proportion of the total cost.

The following example gives a cost estimate for the Westinghouse sulphur cycle. The production capacity of the plant is 100 000 Nm^3/h of H_2.

- HTR with secondary loop	$ 280 million
- electrolysis plant	$ 40 million
- sulphuric acid decomposition	$ 100 million
total capital cost	$ 420 million

The operating costs amount to $ 75 million per annum. The energy produced in the form of hydrogen is 10.2 PJ/year, which gives $ 7.3/GJ as the operating cost component for producing the hydrogen.

The overall efficiency of the process is roughly 40%. This gives the following breakdown for the hydrogen production cost:

energy cost component (0.42/0.40)	$ 1.1/GJ
operating cost component	$ 7.3/GJ
total cost	$ 8.4/GJ

The difference in price between thermochemical hydrogen and hydrogen from future electrolysis is small enough ($ 8.4/GJ as against $ 7.8/GJ) to lie within the limits of error.

Because of the higher efficiency of thermochemical processes, the effect of rising primary energy prices is less than in the case of electrolysis. If the energy cost doubles, the hydrogen price increases to $ 9.4/GJ. If the energy cost triples, the hydrogen price becomes $10.5/GJ. Electrolysis and thermochemical processes can be said to have similar cost structures because they both use nuclear power as primary energy:

- small primary energy cost component
- very high operating cost component.

The nuclear part by itself has operating costs of $ 3.5-4.5/GJ - costs that are already large.

9. Hydrogen from solar energy

The production of hydrogen using solar energy is still at a very early stage of development. As yet there are no experimental plants, so we are unable to give any cost analysis.

Since solar radiation is used, the primary energy cost component is always zero: all costs are capital costs and operating costs.

The cost of solar collectors is still very high, so that industrial use still lies well into the future. As an example we should like to mention the capital costs of a 5 MW_{th} solar furnace at Albuquerque (New Mexico, USA). The plant cost $ 20 million to build, which represents the enormous figure of $ 4000/kW_{th}! Obviously this plant, like the EEC's EURELIOS project (Sicily, Italy), is a prototype test plant; an industrial solar power station will certainly be cheaper.

However, the question remains unanswered as to whether it will be technically possible to reduce these costs to an economically feasible level. This would be around $ 200/kW_{th} (1979). Taking into account that a solar plant will be available for only 2000 hours per year, we arrive at $ 4.2/GJ as the price for the thermal energy obtained.

10. Summary

Table 1 summarizes the results of our analysis of hydrogen production costs. The following conclusions can be drawn from this:

- At the moment, the steam reforming of natural gas and the partial oxidation of heavy oils are the most economic processes for producing hydrogen. The price of hydrogen in this case is $ 5.1-5.7/GJ.

- Hydrogen produced by coal gasification is at present somewhat more expensive but will rapidly become competitive as primary energy prices rise. With a doubling of the primary energy price, coal gasification plants could already produce hydrogen most cheaply.

- The use of nuclear power to produce hydrogen has long had good prospects of becoming economically competitive. The use of nuclear power requires high capital investment as against a very small primary energy cost component. Hydrogen produced by nuclear power could become economically competitive in the future if primary energy prices rise by a factor of 5.

- Electrolytically produced hydrogen has a future only if existing processes can be considerably improved. A substantial reduction of the capital costs is essential.

- The prospects for thermochemical processes are the same as for electrolysis. Both processes are still at an early stage of experimentation, which makes cost predictions very difficult and therefore imprecise.

11. Hydrogen as energy vector of the future

If hydrogen is to be used in the future as an energy vector, it will have to be competitive in comparison with existing energy vectors such as natural gas and SNG (synthesis gas). Table 2 compares the prices of three gaseous energy vectors. The table shows that producing SNG from coal is approx. 15% more expensive than producing hydrogen from coal. The costs shown for "nuclear hydrogen" refer to the average costs for future electrolysis and thermochemical hybrid processes.

The table shows that today's energy vectors can maintain their position even in the future as long as energy prices do not rise by a factor of 5. It is improbable that all energy prices will always increase uniformly. It is possible to imagine a situation in which the price of petroleum has increased by a factor of 5 whilst the price of coal has increased only by a factor of 3, for example. Such price changes influence the competitiveness of energy vectors.

To conclude, we should like to draw attention to movements in primary energy prices and capital costs over the last 10 years. Price movements during the period from 1970 to 1980 were investigated for the following items:

a. petroleum,
b. uranium,
c. Chemical Engineering Plant Cost Index.

Figure 3 shows the increase in costs during the last 10 years (taking 1969 as the baseline). The Chemical Engineering Plant Cost Index is a reliable measure of plant costs in the chemical industry in the USA. The figure shows that oil prices have risen by a factor of 13 in the last 10 years, uranium prices by a factor of 7 and plant costs by a factor of somewhat more than 2.

From this it follows that primary energy costs may rise by a factor of 5 even in the very near future. In looking back over the last 10 years it can be said that the extraordinary rise in oil prices was greatly

Primary energy cost	Natural gas	Synthesis gas (SNG)	Hydrogen (coal)	Hydrogen (nuclear)
1979 level	3.0	6.9	6.0	8.1
2 x (1979 level)	6.0	9.8	8.5	9.3
3 x (1979 level)	9.0	12.7	11.0	10.6
5 x (1979 level)	15.0	18.4	16.0	13.0

Table 2 Estimated costs for gaseous energy vectors (figures in (1979) $/GJ)

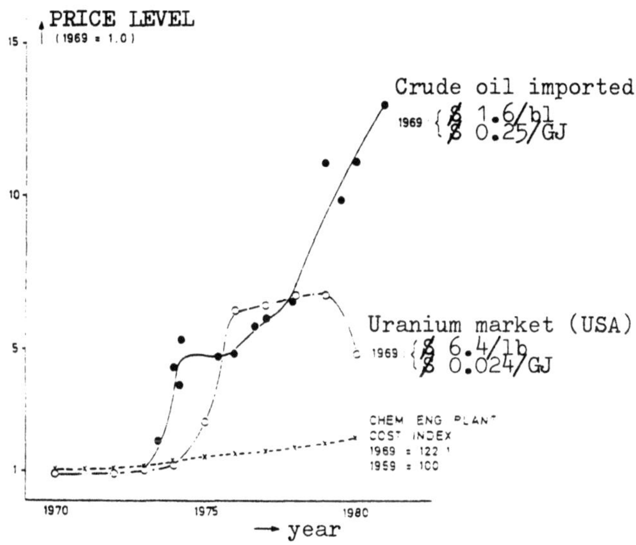

Figure 3 Movements in energy prices

influenced by political factors. However, no one can predict what the next 10 years hold in store for us in this sector.

References

(1) Anon., Assessment of Potential Future Markets for the Production of Hydrogen from Water, Consolidated Final Report, IEA/Commission of the European Communities, Joint Research Centre, September 1980

(2) K.R. Ekman: Production Cost Comparisons of Hydrogen from Fossil and Nuclear Fuels and Water Decomposition, in Proc. of the DOE Thermal and Chemical Storage Annual Contractors Review Meeting McLean, Virginia, October 1980, p 266

(3) Final Report of the Ad Hoc Working Group in view of the Techno-Economical Evaluation of Processes

for Water Decomposition; Commission of the European Communities, Joint Research Centre, Doc. EUR/C-IS-145-78e, March, 1978

(4) J.H. Russell: An Update of Solid Polymer Electrolyte Water Electrolysis Programs at General Electric, Hydrogen Energy Progress, Proc. 3rd World Hydrogen Energy Conference, Tokyo, Japan, 23-26 June 1980, Vol 1, p 3

Trends in the development of hydrogen-powered vehicles using hydride storage

by H. Buchner[1]

1. Introduction

Work done in recent years at Daimler-Benz in the field of hydrogen propulsion has produced the following results. In principle, internal combustion engines can be converted to hydrogen. With a compression ratio of 9:1 and water injection into the cylinder, such an engine would achieve a specific output of some 25 kW/l. The engines can be operated with a greatly reduced fuel ratio ($\lambda = 2.5$, $\emptyset = 0.4$), with loss of power, practically eliminating harmful emissions. In this case the exhaust gas consists mainly of steam.

Safe storage of the hydrogen in the vehicle is made possible by the use of metal hydrides. Because of their high weight of metal per unit volume (compared with carbon), metal hydrides will always be 10-20 times heavier than a tank of hydrocarbon fuel of the same energy content, but they are also 5-10 times lighter than lead batteries for electric vehicles; it therefore follows that hydrogen propulsion will be most useful either where pollutant emissions have to be kept to a minimum (urban and short-distance vehicles, mine vehicles, fork-lift trucks and motor boats), or where hydrogen is to be used as a supplementary fuel in a gasoline/hydrogen mixed engine to make the available gasoline resources go further.

Because of the chemical bonding of hydrogen in metals, the hydrogen reaction always involves heat of reaction. Hydrides therefore store not only hydrogen but also

---

1. Dr Helmut Buchner, Daimler-Benz Aktiengesellschaft, Stuttgart 60

heat. Because of this combination of heat and hydrogen it is possible that, in addition to acting as a fuel tank, hydride storage units may also be able to provide a means of

- cooling,
- supplementary heating,
- water condensation (for injection into the engine) and
- (waste) heat recovery.

Since the construction of a dedicated hydrogen infrastructure would involve very considerable technical and financial burdens, the hydrogen supply would best be organized on the basis of the existing energy infrastructure (natural gas, town gas, electricity). If hydrogen is produced locally from water and electricity or natural gas at the point of use, the process heat arising can be re-used for heating purposes. By using the process heat in producing hydrogen and charging up the hydride storage unit, it is possible to recover part of the energy normally given off as waste heat (eg of the engine) to the atmosphere. Vehicles with combustion engines and hydrogen-hydride storage units are also of interest in connection with nuclear power stations, as they can use some of the nuclear electricity off-peak energy as a gasoline substitute in the electricity/hydrogen conversion.

2. Possible forms of hydrogen storage

If hydrogen is to be reversibly stored for stationary or mobile use, the following technical options are available:

- storage in compressed gas containers;
- storage in liquid form;
- storage in chemically bound form (metal hydrides).

At present, the storage, stockpiling and distribution of large quantities of hydrogen in compressed gas containers and in liquid form is state of the art. The problems involved, such as

- high pressure,
- large volume,
- considerable weight and
- high safety risk

in the case of compressed gas storage, and

- high consumption of primary energy,
- extremely low temperatures,
- high vaporization rates and
- high safety risk

in the case of liquid hydrogen, are offset by the fact that relatively small quantities of hydrogen are used, for example in the case of mobile use in a vehicle.

For this reason, international research has been under way for years with the aim of developing a safe and compact system of hydrogen storage. Work done by the Brookhaven National Labs., International Nickel (INCO), Allied Chemical, Air Products, the Denver Research Institute, KFA/Jülich and the Battelle Institut in Geneva (1-7) has shown that certain metals and their alloys can reversibly store hydrogen chemically or electrochemically by forming metal hydrides. The technical possibilities offered by metal hydrides will be described and discussed in detail below.

3. Survey of metal hydrides

The formation of various exothermic hydrides depending on pressure and temperature and their stoichiometry can be determined by means of concentration/pressure isotherms. If gaseous hydrogen is to react with an alloy, the molecules first have to dissociate into hydrogen atoms which can then be absorbed by the metal. For physical reasons, hydride formation always takes place at constant pressure.

The dissociation of the molecular hydrogen into atomic hydrogen and the bonding in the metal phases means that

- the gas pressure of the hydrogen above the hydride is one to three orders of magnitude below the values for gas storage in high-pressure cylinders,

- the hydrogen density in the hydride is greater than that of hydrogen in liquid form, and
- because of hydrogen's calorific value of 33 000 Wh/kg and the weights of the metal alloys used for storage, the energy densities of the low-temperature hydrides can be about 600 Wh/kg ($TiFeH_2$, $TiFe_{0.8}Mn_{0.2}H_2$) and those of the high-temperature hydrides between 1200 Wh/kg (Mg_2NiH_4) and 2400 Wh/kg (MgH_2).

In practice, energy densities of from 500 Wh/kg (already achieved) to a maximum of 1000 Wh/kg can be attained.

If this is compared with the energy density of gasoline (including tank), ie some 10 000 Wh/kg, and that of a lead battery, ie 25-30 Wh/kg (account being taken of the efficiency of electric propulsion, which is 2-3 times higher than that of the combustion engine), as a rough approximation it can be said that for the same energy content a hydride storage unit (with cladding) will always be 10-20 times heavier than a gasoline tank but 5-10 times lighter than a conventional lead battery with the same range. Thus, reversible chemical storage of hydrogen in metals gives technically acceptable figures in terms of weight, volume, pressure, safety and energy consumption (Figure 1).

The chemical reaction equation for exothermic formation of the hydride from hydrogen and metal is:

$$H_2 + Me \underset{\text{discharge}}{\overset{\text{charge}}{\rightleftharpoons}} \text{hydride} + \text{heat} \begin{array}{l} \nearrow \text{free} \\ \nwarrow \text{added} \end{array}$$

This means that during charging up with hydrogen heat is always produced at the same time, and in order to withdraw the hydrogen from the hydride it is always necessary to add heat.

For the technical application of hydrides the temperature at which their dissociation pressures attain values above 1 bar is of particular interest. Of the many that exist, this group of hydrides (Figure 2, top right) are called low-temperature hydrides (LTH), because even at temperatures below freezing point their dissociation pressure is above 1 bar; all the other hydrides, for which the pressure limit of 1 bar is above the boiling point of water, are known as high-temperature hydrides (HTH) (they can be seen on the left side of the graph).

If the reaction equation for hydride formation is viewed not as a hydrogen reaction but as a thermal reaction, the basic equation for the storage of heat in hydrides is:

$$Q + MeH_2 \rightarrow Me + H_2 \ .$$

Heat is stored and hydrogen is given off by the dissociation of the hydrides.

The back reaction is accordingly:

$$H_2 + Me \rightarrow MeH_2 + Q \ .$$

Now the hydrogen is absorbed and (the formerly stored) heat released. This means that metal hydrides fulfil a dual role. They store both fuel (hydrogen) and heat.

Low-temperature hydrides (LTH) have a heat storage density of up to 0.3 MJ/kg ($TiFeH_2$, $CaNi_5H_6$) or 1.5×10^3 MJ/m^3 in a temperature range from -20°C to 100°C at a hydrogen pressure of ~10 bar.

High-temperature hydrides (HTH) have a heat density of up to 3 MJ/kg (MgH_2, Mg_2NiH_4) or 6×10^3 MJ/M^3 in a temperature range from 150°C to 550°C (TiH_2: 3 MJ/kg, 20×10^3 MJ/m^3, T = 500°C) and at a hydrogen pressure of ~10 bar.

Thus any combustion process (in industry, the home or vehicles) which uses hydrogen from a hydride storage unit can be combined with partial or total storage of the waste heat of combustion of the hydrogen.

52 kg gasoline	18 kg H_2
65 l	200 000 l
Gas cylinder	1000 l, 1400 kg, 200 bar
Liquid (with container)	430 l, 180 kg, 1-2 bar
	Energy -250°C

Metal hydrides	without container	p_{H_2}	with container stage of development
MgH_2	150 l, 260 kg	10^{-2} bar	400 kg, 530 l
"$TiFeH_2$"	220 l, 1000 kg	1-5 bar	1250 kg, 470 l
"$TiCrMnH_x$"	190 l, 850 kg	1-5 bar	1050 kg, 400 l

Lead battery	> 6 t, 3000 l

Figure 1 Energy densities

This dual role of the hydride storage unit allows the following practical fuel/heat coupling mechanism (Figure 3a,b).

In combustion processes (eg H_2 engines or H_2 domestic heating, etc) waste heat is produced and delivered to the hydride to release hydrogen; it is therefore stored in the metal and not given off into the atmosphere simultaneously with combustion. To ensure continuous release of hydrogen, the waste heat from the combustion process has, of course, to be greater than the energy needed for the release of the hydrogen from the hydride.

If the quantity of heat available, together with all the transmission losses, is the same as the energy needed for the dissociation of the hydride, the combustion process produces no waste heat to the outside, since it is then a closed system of hydrogen combustion/waste heat storage. Only when recharging the metal with hydrogen (and only then) does the hydride formation lead to the release of the previously stored heat of combustion. The waste-heat energy can be

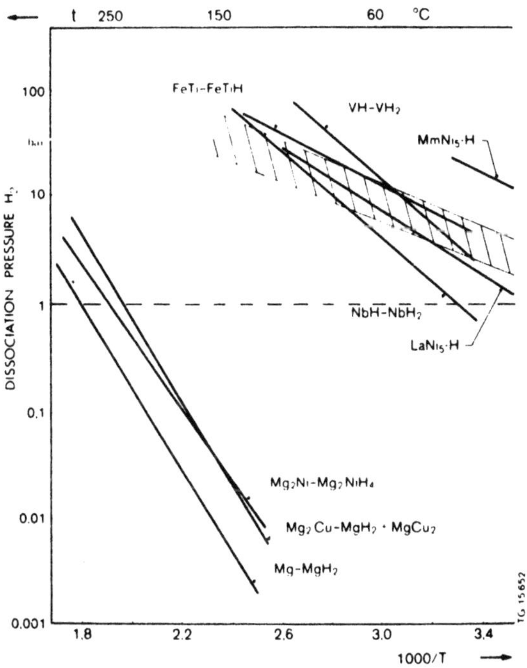

Figure 2 Hydride isochores

made available for practical use, eg for space heating, in various temperature ranges (by varying the pressure in recharging).

This interaction of fuel and heat storage can be used to particular advantage in connection with combustion engines.

4. Selection of hydride systems

4.1 Low-temperature hydrides

The development of hydrides over the last five years (8-12) has led to the following picture for low-temperature storage units (based on TiFe, TiCrMn, $MmNi_5$, etc):

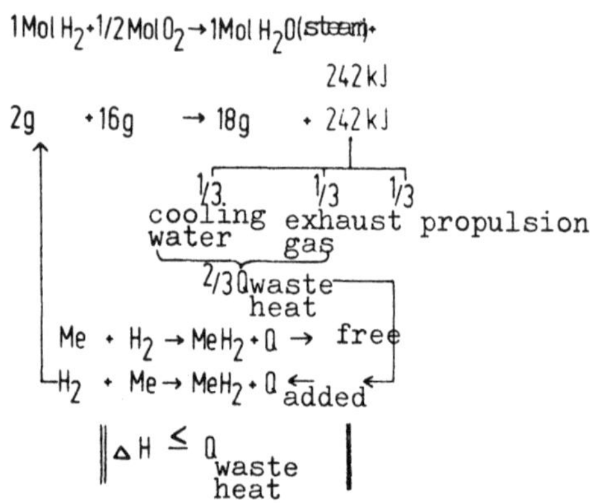

Figure 3a Reaction diagram

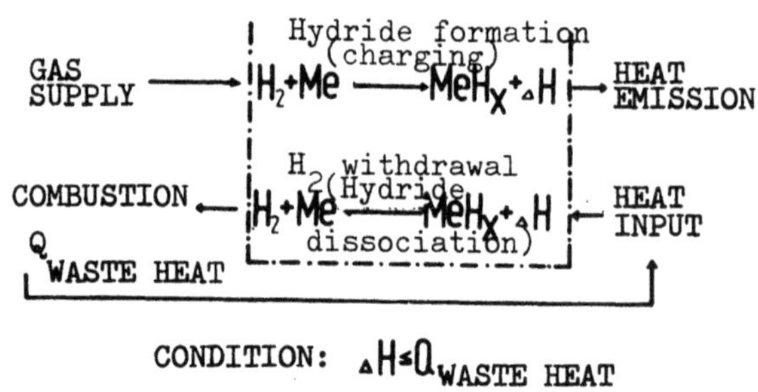

Figure 3b Hydrogen/heat reaction diagram

- up to 2 wt.% H_2 can be reversibly stored, giving an energy density of 660 Wh/kg relative to the weight of the hydride;
- the weight of the hydride container, which consists of tube bundles (Figure 4a,b,c), is about 15% of the weight of the hydride for pressures up to 15 bar, and about 25% for pressures up to 50 bar. The energy density of the storage unit thus reaches values between 500 and 550 Wh/kg.

4.2 High-temperature hydrides as combination units

The high weight factor of hydride storage units in comparison with gasoline can be substantially reduced only by greatly increasing the hydrogen storage capacity, which means only by using high-temperature hydrides.

Since the container weight for high-temperature hydrides is about 40% of the hydride weight in the case of Mg_2Ni hydride and about 50% in the case of Mg hydride, the storage capacity of Mg_2NiH_4 (~1250 Wh/kg) falls to 750 Wh/kg and that of MgH_2 (~2400 Wh/kg) to 1200 Wh/kg.

The proportion of the total formed by the high-temperature unit depends on the quantity of heat between 300 and 350°C available in the engine exhaust gas. It can be shown that, in various town journeys with low exhaust temperatures, about 10% of the hydrogen consumed comes from high-temperature units, while in running tests with higher exhaust temperatures

Figure 4a,b,c

Storage assembly/unit

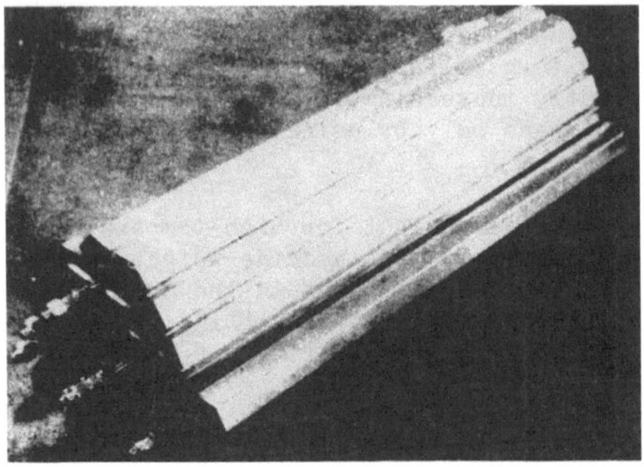

Figure 4b

Figure 4c

(between 400 and 700°C) up to 50% of the hydrogen consumed can be extracted from high-temperature hydrides, and the remaining 50% from low-temperature hydrides. In the first case the energy storage densities are calculated at 525-570 Wh/kg (for the same range a weight factor of 16-18 compared with a gasoline tank), and in the second case of the most favourable running conditions the figure is 625-850 Wh/kg (in this case a weight factor of 11-15 compared with a gasoline tank).

The decisive influence of the container weight on the overall energy density can be seen most clearly in the example of the high-temperature hydride unit. Apart from hydride development, considerable importance therefore attaches also to container development (design, materials, wall thickness, etc). The aim here is to achieve weights of about 10% of the hydride weight in the case of low-temperature hydrides, and about 30% in the case of high-temperature hydrides.

4.3 High-temperature hydrides with supplementary heating

The high energy densities of HT storage units (with no LT hydrides connected) can be fully utilized only when, in addition to the useful exhaust gas energy available, a supplementary heat input is used to cover the energy balance (eg from a supplementary burner directly by exhaust gas heating or indirectly by means of high-temperature oils). Any known fuels (oil, gasoline, methanol, etc) can be used to produce the high-temperature heat needed.

This type of storage system has a number of advantages for special applications:

- Heating of the storage unit is not dependent on petroleum products.
- Continuous combustion in the storage unit heating system gives particularly favourable consumption and emission values.
- At ambient temperature there is a vacuum in the storage unit. Hence the system is absolutely safe in

the event of a leak.

- Recharging of the unit can be done at low pressures (eg 1 bar). Thus neither special safety precautions nor special piping systems are needed.
- The heat released during charging of the system is given off exclusively at a high temperature level.

To verify these observations a separately heated high-temperature storage unit is at present being built at Daimler-Benz for laboratory tests. The problems associated with safety, control, the burner, the heat exchanger and the construction work involved will be examined at the same time (Figure 5).

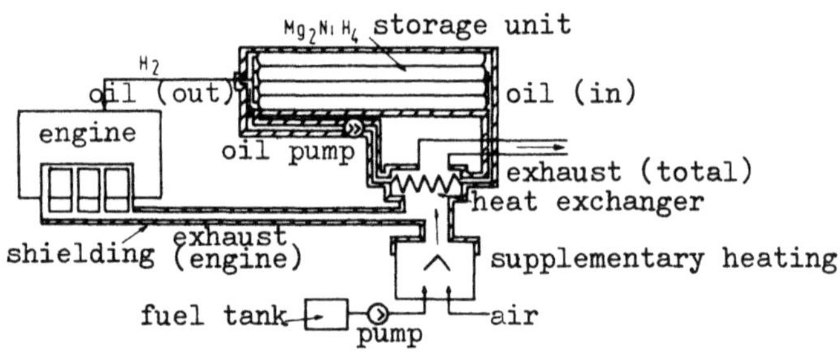

Figure 5 Hydride propulsion with supplementary heat input

Apart from maximum energy density, the aim of hydride development is also to use as much waste heat as possible to release hydrogen, both to make the combustion processes non-polluting (combustion product: water) and to reduce the waste heat substantially. In addition there is the possibility of using for heating purposes the heat given off when charging the unit with hydrogen, and thus saving large quantities of primary energy.

Figure 6 Hydride experimental vehicle (City bus)

Figure 7 Construction and operation of the hydride combination storage unit

1. High-temperature hydride storage unit heated by exhaust gas (can also be used for heating when stationary)
2. Low-temperature hydride storage unit heated by exhaust gas (for further cooling of the exhaust gas - water condensation)
3. Low-temperature hydride storage unit with liquid heat exchanger (can also be used as cooling unit)

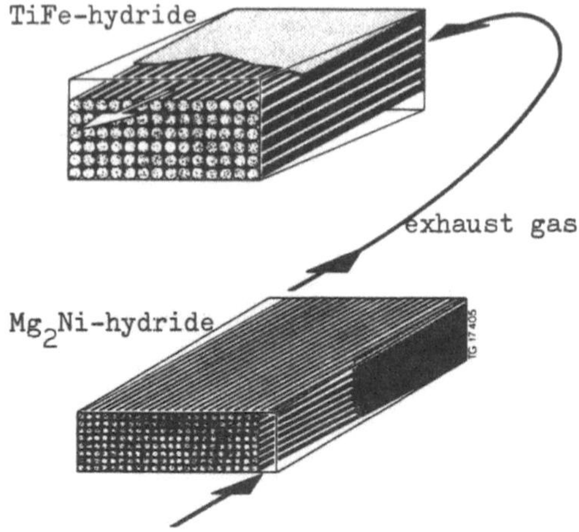

Figure 8 Hydride combination storage unit

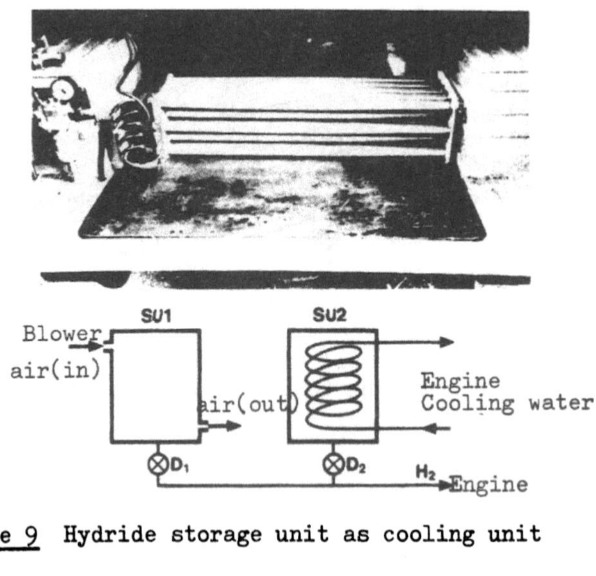

Figure 9 Hydride storage unit as cooling unit

5. State of development (13,14)

The TiFe- and Mg_2Ni-hydride storage units tested at Daimler-Benz in vehicles over several years (Figures 6 and 7) have shown that by appropriate combinations of units with different hydrides (operating at different temperature levels and different pressures) the hydrogen tank can fulfil the following functions:

- Fuel container (consisting for instance of a TiFe hydride unit with a total weight of 220-230 kg and a Mg_2Ni high-temperature unit with a total weight of some 95 kg) having an energy content corresponding to a 20-litre gasoline tank for a 44- to 50-kW engine (Figure 8).

- Air-conditioning unit (Figure 9), with the interior of the vehicle cooled without extra weight or additional fuel consumption (heat from the air being used to release hydrogen from the hydride, at the same time cooling the air to -10°C).

- Stationary heating (Figure 10) of the vehicle, based on the fact that hydrides at different pressures and with different hydrogen bond energies can, on pressure equalization, deliver the difference in the bond energies in the form of heat to the environment.

- Waste heat storage and return during the charging process using a combination of different hydrides. In the extreme case it is possible to develop hydride systems whose H_2 bond energies are so high that all the waste heat from the engine exhaust is needed for H_2 release. In this case the vehicle would not give off any exhaust heat to the environment while moving, which is particularly important in underground travel, eg in mines.

- Water condensation:
  If high- and low-temperature hydrides in the vehicle are supplied with waste heat from the exhaust, the exhaust temperatures behind the combination hydride units may be 50°C, as a large part of the exhaust energy is needed for the release of hydrogen. This

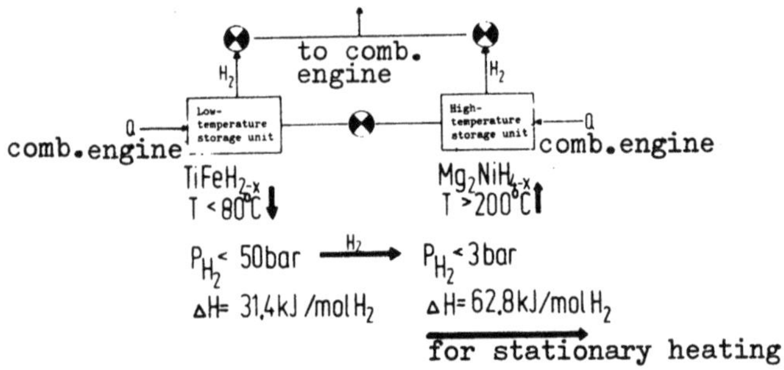

Figure 10 Hydride stationary heating

means that, in the cooling of the exhaust gas during the release of hydrogen from the storage units, complete condensation of the exhaust gas to water also takes place. Part of this water can be re-injected into the cylinder to prevent pre-ignition and backfiring.

6. Demonstration projects of hydrogen vehicles with hydride storage units (15)

To investigate the possible uses and economics of hydrogen systems based on various infrastructures already in existence, a research project is in preparation to be run in Berlin and Stuttgart. The test fleet in Berlin will consist of 20 H_2 vehicles based on Type 230 T and eight passenger vehicles with mixed gasoline/hydrogen operation based on the 280 E. Further vehicles will also be used in Stuttgart. The 20 H_2 vehicles will be fitted with a 2.3-litre internal combustion engine. This engine at present delivers some 50 kW rated output at a compression ratio $\epsilon = 9$ when operating with hydrogen. To prevent backfiring and pre-ignition, water also is injected into the intake. To meet the strict safety and reliability requirements for the customer, a low-temperature storage unit was chosen for the Berlin vehicles, as this is the technology with which most experience has been gained to date. Combined storage systems are to be tested in two small transporters located in Stuttgart.

In mixed gasoline/hydrogen operation, much lower emission of pollutants can be combined with partial substitution of gasoline, without having to forego the latter's favourable energy storage density - in the interests of greater travelling range. Mixed hydrogen/gasoline operation is also favourable for a smooth transition as it is fundamentally possible to burn hydrogen and gasoline in any proportion in the engine.

The passenger vehicle 2.8-litre injection-engine design provides for the engine to be operated with a sometimes very weak hydrogen/air mixture when idling and in the low load range; hence the pollutant emission (eg at road junctions) can be greatly reduced. In the other load ranges the engine is operated with a declining proportion of hydrogen down to a pure gasoline/air mixture. Energy savings of up to 20% are achieved in the Europe cycle and below 10% in the other operating conditions. For the demonstration test carried out in conjunction with the University of Kaiserslautern (Prof. May) (16), there are 11 Type 280 E cars with six-cylinder engines. In Berlin, only low-temperature hydrides will be used as storage units. The total storage unit weight is 260 kg, of which 200 kg is hydride; 3.6 kg of hydrogen is stored. With the consumption figures achieved so far, the travelling range in the CVS test (60% H_2) is 139 km, and that in the EFZ test (70% H_2) is 151 km. Daily travelling distances below 75 km can therefore be achieved with a storage unit weight of some 100 kg. Under these conditions, less than 2 kg of hydrogen per day per vehicle would need to be available. Since no separate infrastructure is being set up for such small quantities of hydrogen - precisely because they are so small - the question immediately arises of where hydrogen can be obtained at present at least to allow mixed gasoline/hydrogen operation.

7. Considerations on the hydrogen infrastructure (17,18)

Technically, the establishment of an extensive hydrogen infrastructure is certainly not feasible in the short and medium term. In order to provide hydrogen in the short term for motor vehicle operation, it will be necessary to have recourse to existing energy infrastructures. In principle any of the existing infrastructures for electricity, natural gas and coal gas, ie energy vectors not used previously for vehicle propulsion, can be utilized to produce hydrogen at the point of consumption. The technical requirements for the production of hydrogen from water by electrolysis (electricity) or from natural gas by separation and from coal-derived gas by pressure swing, are well known; theoretically, therefore, any household which is connected to gas and/or electricity supplies already has the prerequisites for producing its own hydrogen.

In the case of electrolysis the waste heat from the electrolyser, which is generally at about 70°C, can be used for heating. Thus it is possible to use electrolysers which have only a low efficiency. Since the waste heat from the engine that is stored in the hydride tank (about 25-30% of the energy content of the hydrogen) can also be re-used for heating purposes when recharging a vehicle, this represents a relatively favourable opportunity to apply hydrogen technology without extensive changes to the energy infrastructure.

By linking vehicle propulsion and domestic heating it is possible to utilize altogether up to 30-50% of the energy content of the fuel for propulsion instead of the 20% hitherto achieved. The findings made with the small filling station in Stuttgart also provide information on the quantities of heat arising and the time and temperature level at which they are available. These data form the basis of an accompanying study aimed at the proper integration of this heat into the thermal economy of a house. This study is intended to show the technical and economic limits for a combined house/vehicle system, also assuming a pure water supply (Figure 11).

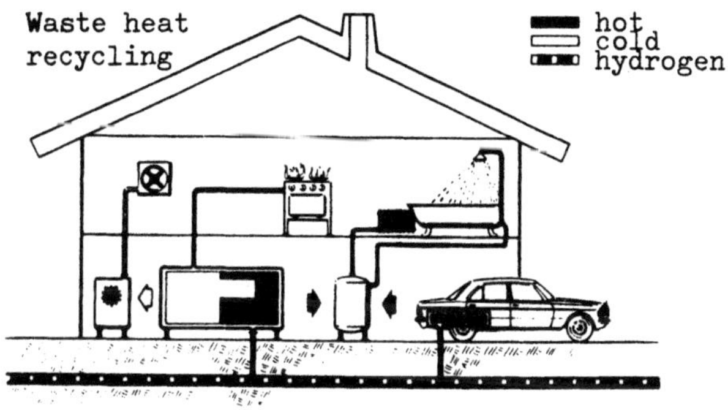

Figure 11 Alternative energies
Hydrogen energy concept

8. Berlin filling station

The hydrogen filling station in Berlin is planned for the site of the Berlin Gas Works (GASAG). The town gas (composition: 53-54 vol.% H_2, 25-26 vol.% CH_4, 18 vol.% CO_2 and 2.5 vol.% CO) will be converted to 99.99% hydrogen by a pressure swing plant. During the day this hydrogen can be used to charge up the hydrogen vehicles in about 10 minutes via 3-4 "quick-acting pumps". The maximum capacity of the filling station is 240 m^3 H_2/h. For rapid charging up there is a charging pressure of 50 bar. The Berlin station is intended not only to demonstrate the speed of re-charging but also to serve as an example showing that the operation of hydrogen vehicles does not necessitate a dedicated infrastructure in the form of hydrogen supply networks, as existing gas distribution networks can be used.

9. Outlook for the future

Work in the field of hydride development and hydrogen production in the last few years has shown that hydrogen as an energy vector can be used safely and economically by means of H_2 storage in metal hydrides but that, in the present state of the art, the large-scale production and distribution of pure hydrogen from water seems favourable in only a few cases. The possibility of making use of existing energy infrastructures for hydrogen production, at the same time utilizing the waste heat produced, therefore seems to provide the key to entry into an era of hydrogen technology. Greater activity in precisely these areas should be a priority objective, in order to determine to what extent hydrogen technology can help to relieve the present or the future energy situation, especially in the vehicles sector.

References

(1) J.J. Reilly and R.H. Wiswall, Jr: The Reaction of Hydrogen with Alloys of Magnesium and Nickel, Inorganic Chem. 7 (1968)

(2) G.D. Sandrock: The Metallurgy and Production of Rechargeable Hydrides, Proc.Int.Symp. on Hydrides for Energy Storage, Geilo, Norway, 1977

(3) G.G. Libbowitz: The Metal Hydrides for Thermal Energy Storage, 9th IECEC, San Francisco, California USA (1974)

(4) G.D. Sandrock: A New Family of Hydrogen Storage Alloys based on the System Nickel-Mischmetal-Calcium, 12th IECEC, Washington DC, USA, 1977

(5) E. Lebsanft, G. Arnold, K.H. Klatt, M.A. Pick, J. Töpler, H. Wenzl: Properties of FeTi-hydrides used for Hydrogen Storage, Proc. 2nd Int.Congress on Hydrogen in Metals, Paris, France, June 1977

(6) C.E.E. Lundin and F.E. Lynch: A Detailed Analysis of the Hydriding Characteristics of $LaNi_5$, Proc. 10th IECEC, Newark DE, USA, 1975

(7) O. de Pous and H.M. Lutz: Effect of the Interstitial Hole Size and Electron Concentration on Complex Metal Hydride Formation, Proc. 2nd World Hydrogen Energy Conference, Zürich, Switzerland (1978)

(8) S. Suda and M. Uchida: Mixing Effects of Different Types of Hydrides, Proc. 2nd World Hydrogen Energy Conference, Zürich, Switzerland (1978)
(9) I. Jacob and D. Shaltiel: The Influence of Al on the Hydrogen Sorption Properties of Intermetallic Compounds, Proc. 2nd World Hydrogen Energy Conf., Zürich, Switzerland (1978)
(10) H. Buchner, M. Stohrer and O. Bernauer: Cubic Metal Alloys for Hydrogen Storage, Proc. 2nd World Hydrogen Energy Conference, Zürich, Switzerland (1978)
(11) F.E. Lynch and E. Snape: The Role of Metal Hydrides in Hydrogen Storage and Utilization, Proc. 2nd World Hydrogen Energy Conf., Zürich, Switzerland (1978)
(12) H. Buchner, O. Bernauer and W. Strauss: Development of High Temperature Hydrides for Vehicular Application, Proc. 2nd World Hydrogen Energy Conf., Zürich, Switzerland (1978)
(13) H. Buchner: The Hydrogen Hydride Energy Concept, Int.J.Hydrogen Energy 3 (1978), 385-406
(14) J. Töpler, O. Bernauer and H. Buchner: The Use of Hydrides in Motor Vehicles, J. of Less Common Metals 74 (1980), 385-399
(15) R. Povel and H. Buchnor: Konzept zur Erprobung der Wasserstoff-Technologie (Concept for testing hydrogen technology), BMFT Status Seminar Reports, November 1980
(16) H. May, U. Hattingen and W. Jordan: Theoretical and Experimental Investigations on Hydrogen-Gasoline Hybrid Operation in I.C. Engines, Proc. 16th Fisita Congress, Tokyo, 1976, Paper 5/19
(17) H. Buchner: Perspectives for Hydrogen Hydride Technology, Progress in Energy and Combustion Science, 1980/4
(18) H. Buchner: Zur Frage der Wasserstoff-Infrastruktur für Kfz (Hydrogen infrastructure for motor vehicles), Chemie-Technik 10 (1981), No 4

Energy storage with hydrogen

by W. Peschka[1]

Introduction

The not too distant exhaustion of naturally occurring oil and gas is forcing the highly industrialized countries to cut down their use for energy production. This situation is exacerbated by the growing energy requirements of the Third World countries and by the dependence of the European countries on imports of energy and raw materials.

The safeguarding of energy supplies demands both the rational use of existing energy resources and the discovery of new sources of energy. From Germany's point of view these can only be coal and nuclear energy in the short and medium term, while in the long term regenerative primary energy will be needed to an increasing extent. Such energy is as a rule not suitable for end use and is mostly available only at a distance from the point of consumption. It must therefore be converted into secondary energy vectors suitable for transport and direct consumption and for balancing energy supply and demand by means of storage.

No single traditional or novel secondary energy vector, such as electricity, remote heating, synthetic hydrocarbons or hydrogen, can satisfactorily meet the total energy requirements of the future. Hydrogen satisfies the basic criteria for a secondary energy vector, and its advantages and disadvantages must therefore be compared with those of the alternatives. It is very important to have in readiness options for secondary

1. Prof. Dr W. Peschka, Institut für Technische Physik, Deutsche Forschungs- und Versuchsanstalt für Luft- und Raumfahrt eV (DFVLR), Stuttgart

energy vectors which are compatible with alternative lines of development, so as to ensure the greatest possible latitude of choice in future energy policy.

Hydrogen represents such an option, since it facilitates a change-over to non-fossil primary energy sources in all conceivable future energy-supply structures, and also permits a continuous transition between different forms of gaseous secondary energy vectors (Figure 1). Non-fossil derived hydrogen will certainly not become a competitive secondary energy vector in the present century. However, during the next few decades decisions must be taken about which path to follow in a future based largely on non-fossil energy. Until then it is necessary to keep open the hydrogen option, ie to solve the problems still remaining connected with its production, distribution and use, and also to demonstrate that hydrogen can be integrated into future energy systems on a scale comparable with present consumption of oil and natural gas.

Compared with electricity hydrogen has excellent storage and transport properties, and can be converted efficiently into other forms of secondary energy without any significant formation of noxious products. At the present time it suffers from the disadvantage of relatively high production costs, especially if it becomes necessary to abandon present methods of making it from oil or natural gas. Hydrogen is used today mainly as a raw material in the chemical and petrochemical industry; its use as a secondary energy vector has so far been restricted to a few special cases (Figure 2).

From the point of view of a compulsory retreat from fossil primary energy sources and the increasing importance of regenerative energy systems hydrogen represents in the long term an essential supplement to existing secondary energy vectors.

Furthermore, the use of fossil-based primary energy cannot possibly be viewed without taking into account the problem of atmospheric carbon dioxide (Figures 3

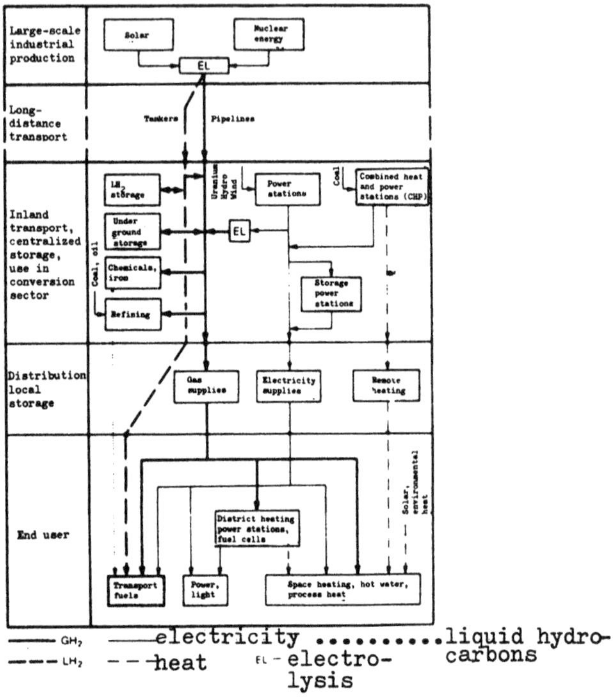

Figure 1 Elements of an energy economy based partly on hydrogen (see (12))

and 4). A gradual transition from fossil energy sources to hydrogen, which is a completely regenerable secondary energy vector, would constitute an effective measure against the rising content of carbon dioxide in the atmosphere. This would make possible an ecologically closed material cycle, which is kept going by the supply of primary energy and which would permit a more intensive global energy turnover than, for example, the carbon cycle.

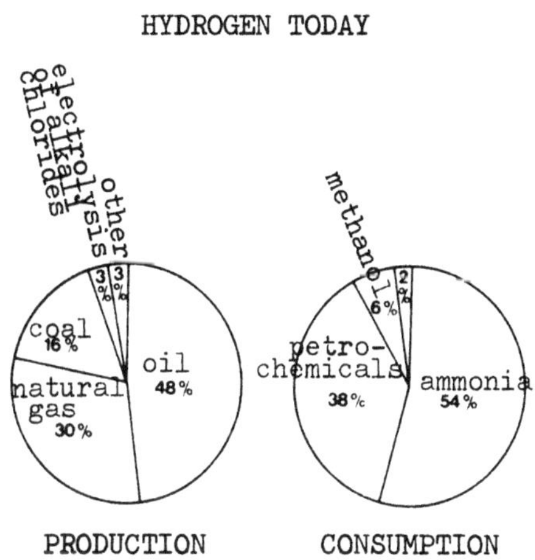

Annual world production 1974:	$220 \times 10^9 m^3 = 88 \times 10^6$ TCE
Share in Federal Republic of Germany:	3%

Figure 2 Survey of current uses of hydrogen

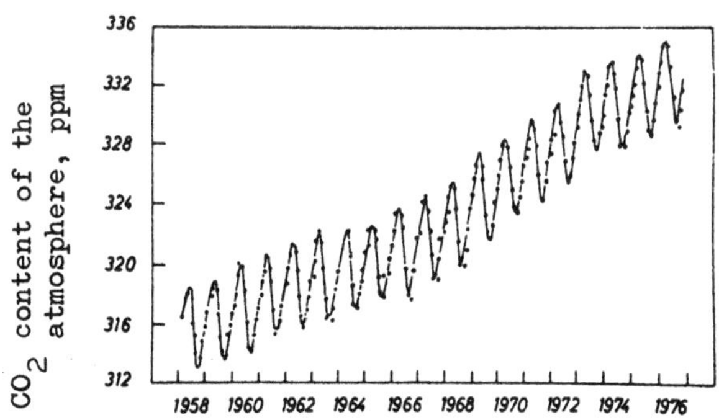

Figure 3 Measurements of the CO_2 content of the Earth's atmosphere at the Mauna-Loa observatory

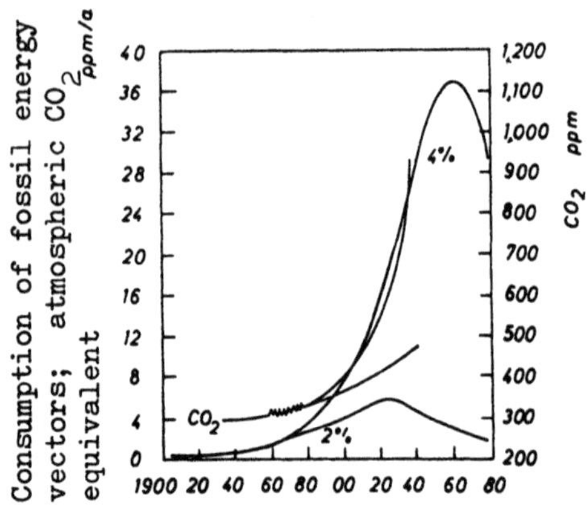

<u>Figure 4</u> CO_2 content of the Earth's atmosphere as a function of the consumption of fossil-based primary energy vectors

<u>Hydrogen storage</u>

Since hydrogen has excellent storage and transport properties compared with electricity, and can be readily and efficiently converted to other forms of secondary energy, energy storage can be advantageously effected by the storage of hydrogen. The problems of energy storage are therefore in part the problems of hydrogen storage.

Hydrogen can best be stored in different ways, depending on the various fields of application.

Storage can be effected in a gaseous or liquid state, or in the structure of solids. According to the field of application it is necessary to distinguish between storage on a large scale and on a small scale. The former applies particularly to stationary applications, and the latter also to mobile ones (Figure 5).

<u>Storage as a gas</u>

Hydrogen gas can be stored at ambient temperatures, or

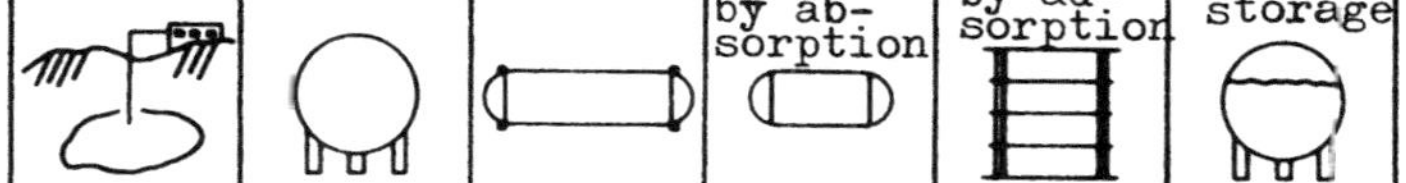

Hydrogen storage	LARGE-SCALE STORAGE (stationary)					
	Storage under pressure			storage by ab-sorption	storage by ad-sorption	LH_2 storage
	under-ground	low pressure	high pressure	metal hydride	cryoad-sorbent	
State of the art	yes	yes	yes	limited	limited	yes
Problems	accessi-bility		safety	resources safety	not yet defined	safety
Economic factors (T = capacity/output, kWh/kW)	$T > 0$	$T < 30$	$T < 30$	$T < 15$	$20 < T < 40$	$T > 40$
Competing methods		under-ground storage	under-ground storage	storage under pressure	LH_2 storage	under-ground storage
Area of use	tied to specific loca-tions	short-term storage	short-term storage	as for storage under pressure	inter-mediate range	• long-term storage • year-long storage

Figure 5a Survey of methods of storing hydrogen

Hydrogen storage	STORAGE OF SMALL QUANTITIES				
	Residential areas, etc		Surface transport		Air transport
	 Metal hydride	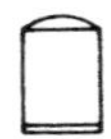 LH_2 storage	 Metal hydride	 LH_2 tank	 LH_2 tank
State of the art	limited	yes	limited	limited	limited
Problems	• resources • dehydrogenation • safety	• improvement of insulation • safety	• resources • heavy weight • safety	• infrastructure • safety	• airport facilities • safety
Economic factors	specific applicability	long-term storage	small travelling range	large travelling range	long-haul transport
Area of use	short-term storage	long-term storage	urban transport	transport of passengers and goods	intercontinental transport

Figure 5b Survey of methods of storing hydrogen

by physical adsorption at around -200°C (cryoadsorption). The low density of hydrogen gas at ambient temperatures is a disadvantage, since it implies bulky storage plant and higher pressures in order to increase the amount that can be stored. This applies also to cryoadsorption.

Underground storage in caverns or pores is basically suitable for hydrogen also, and offers the same economic advantages as its current use for natural gas. Such storage is indispensable for gas supplies in order to even out the seasonal variations in demand. Its technical and economic suitability for hydrogen can be assumed on the basis of current experience in using storage in caverns for town gas containing up to 65% hydrogen at pressures of up to 160 bar.

Similarly, pipelines can be used as short-term storage to balance output with demand, as is done today for natural gas. Intercontinental pipelines would also be able to a large extent to even out the day/night rhythm of hydrogen produced by solar energy, so that continuous withdrawal of gas with only slight pressure variations would be possible.

Storage in medium- or high-pressure vessels can be used basically on a daily or weekly basis. The low density of hydrogen is a disadvantage, since it implies that the amount of energy stored is smaller than for other compressed gases. For this reason compressed hydrogen will make only a very small contribution to the storage of energy in any future hydrogen economy.

Cryoadsorption employs the storage of hydrogen at medium pressures (up to 40 bar) and temperatures around -200°C by physical adsorption on suitable adsorbents (active carbon). It permits the storage of considerable amounts of hydrogen (approximately 7 wt.%) in relatively compact pressure vessels, while avoiding the trouble of liquefaction. This method is at present being investigated on a laboratory scale, but its realization on a large industrial scale requires further development work. This method of storage is likely to be economic only for storage capacities (in terms of output) of about 20-30 kWh/kW, for storage by the week.

According to present knowledge it should be possible to store hydrogen gas economically under pressure when underground storage is available.

Storage as hydrides

Hydrogen can be stored in chemical combination with metals as metal hydrides. The necessary storage volume is comparable with that for storage as a liquid, but the mass is much greater, since the stored hydrogen constitutes only a small proportion of the mass of the hydride. The only compound at present available for use on a large industrial scale is iron titanium hydride. When operated cyclically at pressures of up to about 20 bar this will store about 1.3 wt.% of hydrogen in large-scale plant. This figure is not much larger than the amount of gaseous hydrogen which can be stored under pressure, expressed as a percentage of the weight of the container. Consequently, the expenditure on the metal hydride alloy makes the process more costly than the storage of the gas under pressure. Furthermore, considerable amounts of heat must be removed when the hydride is formed and supplied when the hydrogen is withdrawn.

Hydrides with a higher content of hydrogen either involve expensive or scarce raw materials, such as nickel, or (as with magnesium) there are special problems connected with the operating temperature and the large amounts of heat to be exchanged. Extensive investigations have already been carried out on the physical and chemical properties of metal hydrides, but not enough is yet known about special problems such as long-term operation, contamination of the storage material, etc.

Storage of liquid hydrogen

The storage of hydrogen as a liquid requires cryogenic

storage vessels, but has the advantage over the gas that large amounts can be stored in a small volume. The necessary storage vessels with perlite-vacuum insulation are available commercially. The largest vessels so far constructed were in the NASA Apollo programme; they had a capacity of 3200 m^3, and on account of the low evaporation losses storage could be maintained for several years.

For the same amount of hydrogen or energy stored, liquid hydrogen containers are about 30 times less expensive than hydride storage units or pressure storage vessels. In consequence, for storage capacities (in terms of output) of about 50 kWh/kW upwards liquid storage represents the most economic alternative, since the low cost of the storage tanks, which depends only slightly on their capacity, outweighs higher running costs and higher capital costs, such as those for liquefaction plant. For year-long storage with a typical capacity (in terms of output) of a few hundred kWh/kW liquid storage is the sole option (Figure 6).

Hydrogen in electricity generation

In future energy systems the electricity available for the large-scale production of hydrogen will fluctuate greatly with time. In regenerative energy systems this is a direct consequence of the irregular supply of primary energy. In systems based mainly on nuclear energy, developed to the level of medium loads, electrical energy for the production of hydrogen will at first be available only during periods of low demand. It should be noted that hydrogen production can be introduced advantageously as a means of storing energy even in the initial and transitional phases of any non-fossil energy system.

Nuclear power stations can be operated most efficiently at a constant load level, independently of fluctuations in demand, and the excess output can be stored in the form of hydrogen.

By using hydrogen for storage, regenerative energy systems can contribute to a continuous availability of

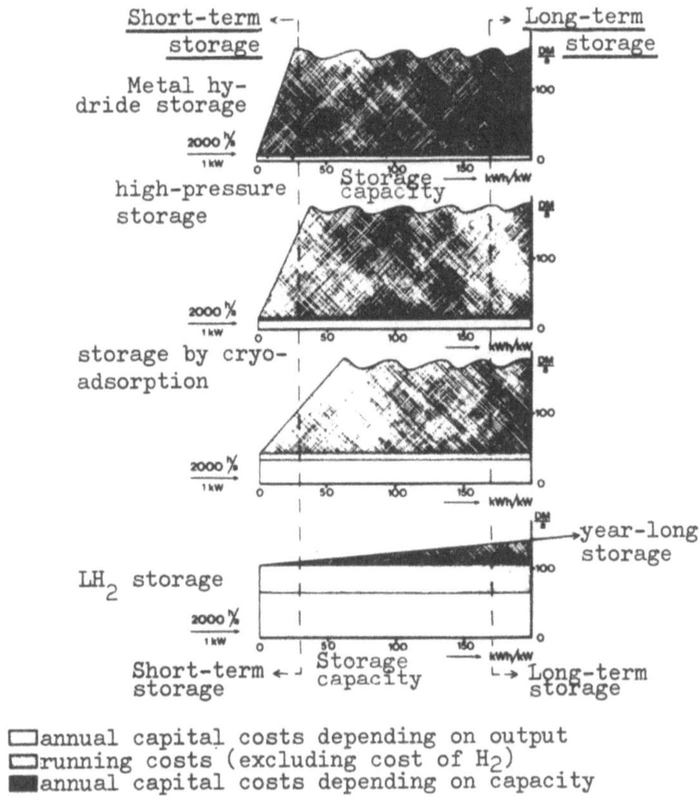

<u>Figure 6</u> **Comparison of the annual costs of different methods of H_2 storage**

energy, independent of diurnal or seasonal fluctuations in the energy supply. Special attention should be drawn to the combination of water electrolysis with solar power stations, which produces hydrogen in two stages of energy conversion, without a detour through thermal and mechanical energy. A precondition for this will be a further substantial decrease in the cost of silicon solar cells. Ways must also be found for raising the enormous capital expenditure needed to erect the large photovoltaic power stations of the future, bearing in mind the associated global and political problems (Figures 7 and 8).

Apart from these applications, which lie rather in the future, hydrogen can provide an advantageous coupling between gas and electricity supply systems. In this way it is possible to run base-load power stations with maximum efficiency at constant output, independently of fluctuations in demand, and to feed the excess output in the form of hydrogen into an existing gas distribution network, or alternatively to use it for producing electricity for peak loads.

It is also possible to store hydrogen (and oxygen) locally and to burn it to produce electricity at periods of peak demand (hydrogen storage power stations). This indirect storage of electricity can supplement pumped-water storage, which is limited by local conditions.

The introduction of hydrogen power stations is not restricted to large centralized installations; it is also possible to set up small decentralized installations with a high conversion efficiency, for example in densely populated areas, not least in virtue of the special environmental suitability of hydrogen.

Application and improvement of established technologies

These include hydrogen-air turbines and hydrogen-oxygen steam generators.

Gas turbines fuelled by oil or natural gas are used in electricity generation for covering peak loads.

Hydrogen offers advantages over gas turbines using fossil fuels on account of the very small proportions of noxious substances in the exhaust gases. Hydrogen-oxygen steam generators are at the experimental stage and can be developed further on the basis of existing experience with the combustion chambers of rocket motors (Figures 8, 10, 11, 12 and 13).

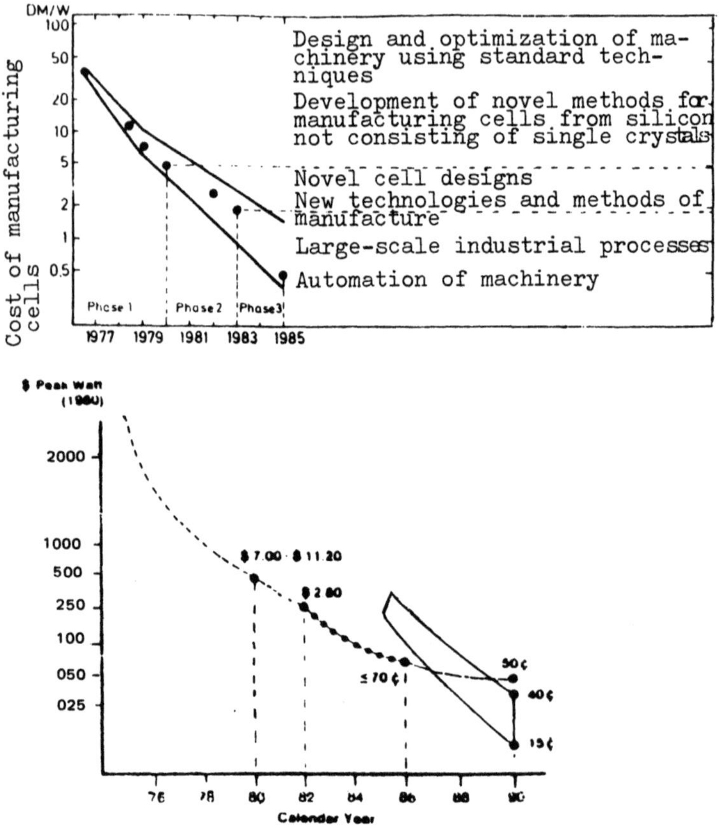

Figure 7 Possible developments in the cost of manufacturing silicon solar cells (see (13))

The application and further development of new technologies

This comprises the production of electricity by using hydrogen in fuel-cell systems.

The modular construction of fuel-cell systems makes it possible to obtain favourable efficiencies even with small assemblies (Figure 9). A further advantage of the modular structure of fuel-cell systems lies in the extremely high potential for reducing manufacturing costs when producing a large number of units. However, in the present state of fuel-cell technology such applications are feasible only for pilot plants. A number of developments are still necessary, especially in the fields of electrodes and catalysts, to make fuel-cell systems applicable on a large scale (Table 1).

It is also of importance that the conversion efficiency of fuel cells is independent of the load factor over a wide range, so that a high efficiency can be obtained even with partial loads. Furthermore, with fuel cells which work at high temperatures high-grade waste heat can be used for thermal energy production.

On the basis of experience of electricity generation with small hydrogen-oxygen assemblies, research and development work is now in progress, especially in the USA, with the object of developing fuel-cell power stations for the centralized and local generation of electricity. The present position with regard to various fuel cells can be described as follows.

Low- and medium-temperature fuel cells with operating temperatures between 80 and 250°C have been equally far developed in three directions:

a. alkaline fuel cells with aqueous caustic potash as electrolyte;

b. acid fuel cells with concentrated phosphoric acid solutions as electrolyte;

c. solid-electrolyte fuel cells with acidic ion exchange groups in a polymer matrix.

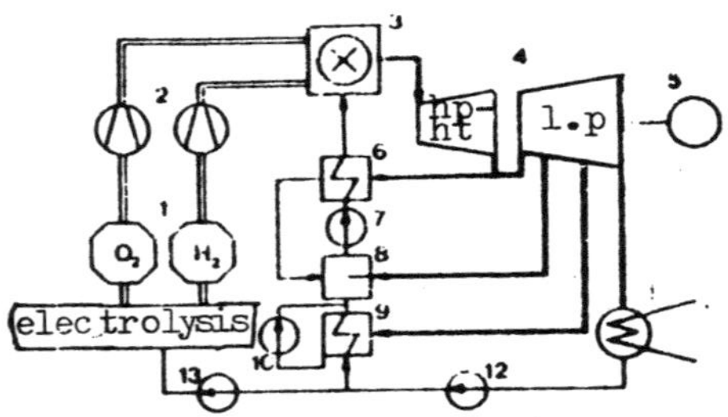

1. storage
2. compressor
3. steam generator
4. turbine
5. generator
6. h.p. preheater
7. feed-water pump
8. degasser
9. l.p.preheater
10. intermediate condensate pump
11. condenser
12. main condensate pump
13. electrolysis condensate pump

H_2/O_2 steam-turbine process

(Type A power station)

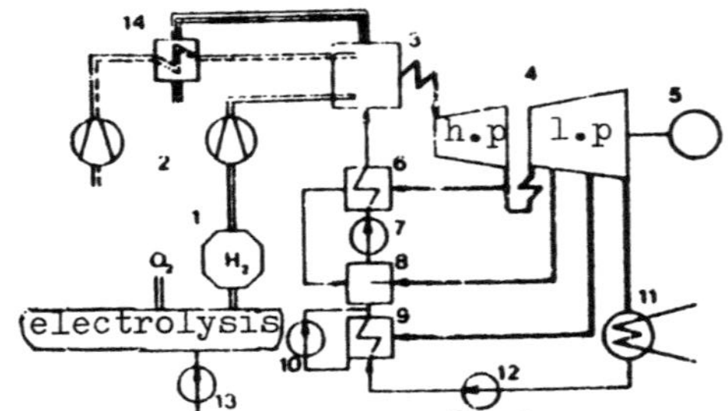

1. storage
2. compressor
3. steam generator (conventional)
4. turbine
5. generator
6. h.p. preheater
7. feed-water pump
8. degasser
9. l.p. preheater
10. intermediate condensate pump
11. condenser
12. main condensate pump
13. electrolysis pump
14. air preheater

Conventional steam-turbine process with H_2-fired boiler

(Type B power station)

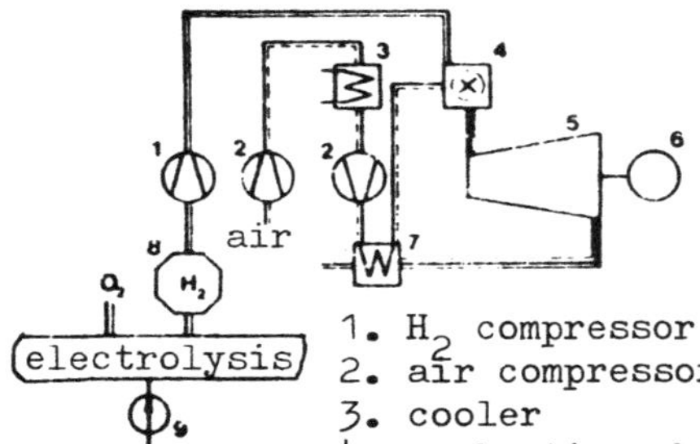

1. H_2 compressor
2. air compressor
3. cooler
4. combustion chamber
5. turbine
6. generator
7. heat exchanger
8. storage
9. electrolysis pump

Open H_2/air gas turbine process with intermediate cooler and recuperator (Type C1 power station)

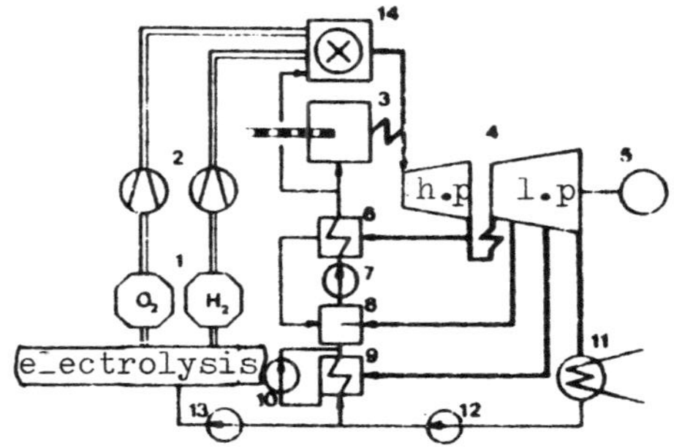

1. storage
2. compressor
3. steam generator (conventional)
4. turbine
5. generator
6. h.p.preheater
7. feed-water pump
8. degasser
9. l.p. preheater
10. intermediate condensate pump
11. condenser
12. main condensate pump
13. electrolysis pump
14. steam generator (H_2/O_2)

Fossil-fired power station for base or medium loads with additional feed of H_2/O_2 steam

Figure 8 Steam-turbine and gas-turbine processes for producing electricity from hydrogen (see (13))

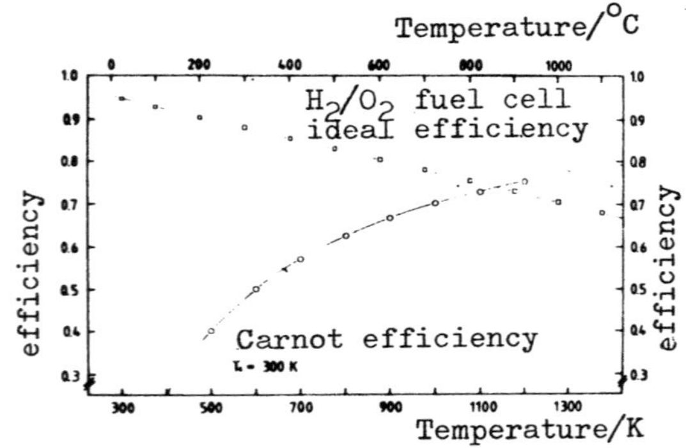

Figure 9 Efficiency of H_2/O_2 fuel cells (see (13))

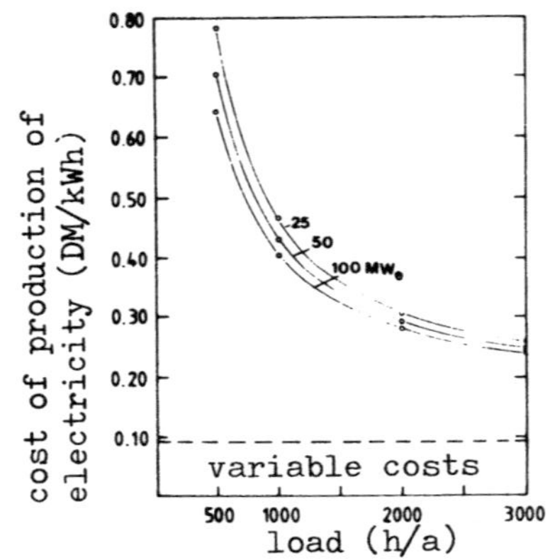

Figure 10 Cost of production of electricity from H_2/O_2 fuel-cell power stations (see (13))

Table 1 Operational data, costs and availability of fuel-cell systems (see (13))

	LOW-TEMPERATURE FUEL CELL (NTBZ)			HIGH-TEMPERATURE FUEL CELLS	
ELECTROLYTE	KOH	H_3PO_4	ion exchange membrane	$(Li,Na,K)_2CO_3/LiAlO_2$ Matrix	$(ZrO_2)_{1-x}(SE_2O_3)_x$
T (°C), P (bar)	80-200, 2-5	150-200, 4-8	75-150, 4-8	650, 5-10	1000, < 2
Fuel	H_2	H_2, H_2/CO	H_2, H_2/CO	H_2/CO	H_2/CO
Catalyst, anode	Ra-Ni/Ni	Pt/C, WC/C	Pt/Nb	Ni	Ni-ZrO_2 CERMET
Catalyst coverage (mg/cm²)	120	1,40	1,3	----	---
Oxidizing agent	O_2	air	O_2, air	air/CO_2	air
Catalyst, cathode	Ra-Ag/Ni	Pt/C	Pt/Nb	$LiNiO_x$/Ni	$(SE,Sr)_x(Mn,Co)O_{3-x}$
Catalyst coverage (mg/cm²)	60	1	1,3	----	----
Density of output* (W/cm²)	0,32	0,20	0,15-0,20	0,13	0,15 - 0,30
Output category**	≤ 50 KW	4,8MW(1980181)	≤ 10 KW	≤ 22 KW	≤ 1 KW
Overall electrical efficiency*	0,52	0,40 - 0,45	0,40 - 0,45	0,45	0,45 - 0,55
Lifetime of system	> 1000 h	> 5000 h	> 5000 h	----	----
Lifetime of individual cells	> 20000 h	> 20000 h	> 30000 h	> 15000 h	> 22000 h
Capital costs*** (DM/kW$_e$)	3000-4000	2000 - 3000	3000 - 4000	> 3000	----
Capital costs**** (DM/kW$_e$)	700-1200	(500)700-1000	(500)- 1000)	700 - 1000	900 - 1500
Period of development	≤ 1990	1982-1985	> 1992	> 1990	> 2000

*realized at stated loads

**systems actually constructed

***DM 1978, present-day technology, excluding R&D costs

****DM 1978, future technology, MW range

Alkaline fuel cells can be used advantageously for producing electricity from pure hydrogen and oxygen, since high efficiencies can be obtained here with cheaper catalysts. Fuel cells with acid electrolytes can also be operated with fossil-derived impure hydrogen or hydrocarbons, but as yet require the use of catalysts of platinum or other noble metals.

Alongside the use of fuel cells operating at low or medium temperatures and pressures, the development of optimum configurations of large fuel-cell modules therefore depends on the choice of suitable inexpensive materials for electrodes of sufficient long-term stability and enhanced procedures for manufacturing the individual components and modules.

High-temperature fuel cells operate in the temperature range from 600 to 1000°C. Because of their high operating temperature they have the advantage that no special catalysts are needed to activate the reactions, so that crude gas or CO/H_2 mixtures can be used directly for producing electricity, and high-grade waste heat is available for use.

High-temperature fuel cells with solid electrolytes based on conduction by oxide ions represent the reverse of the high-temperature electrolysis of steam, and involve the same problems as those described in that connection. In particular, it will be necessary to solve the fundamental problems connected with constructional materials arising from the high operating temperatures (around 1000°C) before development to an industrial scale is possible.

A more advanced stage of development has been reached with high-temperature cells with molten carbonate electrolytes, operating at about 600°C. Laboratory cells have been undergoing tests for some time now. Further development into larger units depends essentially on solving basic constructional problems and developing suitable materials. If these development problems could be solved fuel-cell power stations could compete seriously with H_2/O_2 or H_2/air steam-turbine processes (Figures 10, 11 and 12).

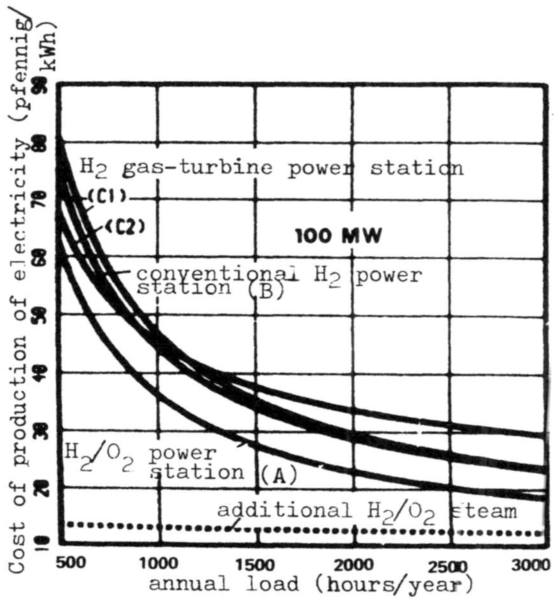

Figure 11 Cost of production of electricity for 100 MW hydrogen power stations with gas and steam turbines (see (13))

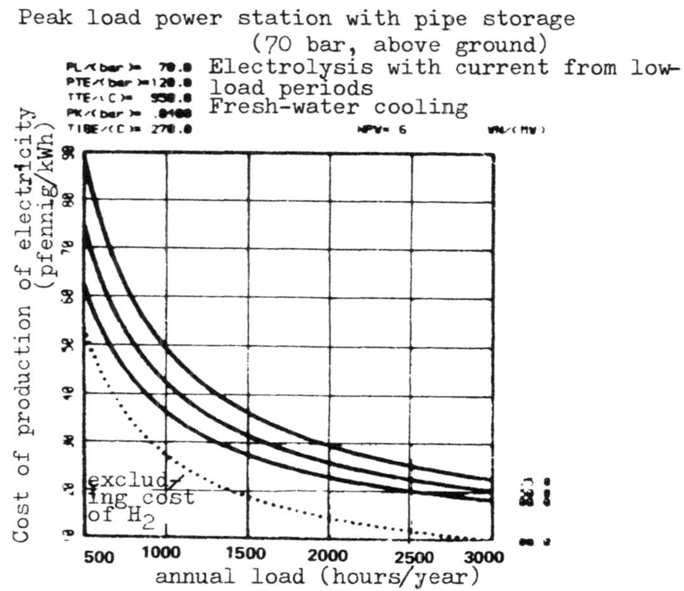

Figure 12 Cost of production of electricity for H_2/O_2 steam-turbine power stations with various nominal outputs (see (13))

Hydrogen as an alternative transport fuel

In the medium to long term hydrogen can be regarded as an alternative to fossil or synthetic hydrocarbons on account of its favourable environmental properties. These properties are shared by electricity, which has some prospects of use on a modest scale for personal surface transport, but none at all for air transport. As many transport fuels become scarcer, recourse will first be had to synthetic fuels from coal or from oil shales and tar sands to meet requirements. However, it should be noted that considerable amounts of carbon dioxide are discharged into the atmosphere not only in the combustion of hydrocarbons but also in the manufacture of synthetic hydrocarbons from coal. The sole exceptions to this are the direct hydrogenation of coal with hydrogen produced by environmentally harmless processes, and the use as transport fuels of ethanol or methanol derived from vegetable carbohydrates (photosynthesis), especially from the residues from the processing of sugar-cane (Brazil). However, the last-mentioned is feasible only in regions with a suitable climate and relatively low transport requirements.

As a candidate among these alternative transport fuels hydrogen has real chances for steadily increasing use, provided that its manufacture by environmentally harmless means (which of late is shared by all synthetic fuels) is supplemented by research and development work on the creation of the necessary infrastructures for its various applications. Primary importance attaches not only to safety aspects and problems of distribution but also to storage on board the vehicle. For road transport storage either as hydrides or as liquid may be considered, according to the circumstances. For rail transport liquid storage holds the most promise, while for air transport hydrogen can be used only as a liquid.

For sea transport the use of liquid hydrogen seems to be indicated, if tankers are required for carrying hydrogen from remote production sites (tropical seas for utilizing solar energy or heat from the ocean, arctic regions for using the energy of glaciers).

Figure 13 View of an H_2/O_2 steam generator for 15 MW thermal output (DFVLR-Institut für Chemische Antriebe und Verfahrenstechnik, Lampoldshausen)

Hydrogen as an alternative fuel for aircraft

Liquid hydrogen has for a long time been without rivals as the future aircraft fuel for the extreme supersonic range. Recently, not least as a result of rising fuel costs (which at present average about 40% of total operating costs), possibilities for its use in the subsonic range are also beginning to appear.

The properties of any alternative transport fuel have to be weighed against those of the fuels in use at present, and compelling arguments are necessary before the introduction of alternatives can be considered. One such argument is certainly the rise in oil prices to be expected in the next century on account of the anticipated drastic decline in oil reserves. However, a global change-over to other aircraft fuels on these grounds alone will certainly not be one of the first expedients adopted under such circumstances, since the associated reorganization of major branches of industry will require enormous capital expenditure. Nevertheless, compared with other potential users air transport seems to be particularly suitable as a pioneer and large-scale user of hydrogen on account of its international ramifications, the logistic possibilities of present-day airports and its ability to push through major management projects for technical developments on a large scale in a relatively short time.

Since the development of new types of aircraft takes about ten years, and their life-span is about twenty years, from the point of view of economics today's planning for the next generation of aircraft already has to take into account the fuel costs to be expected in the 1990s. Since the next generation of aircraft is thus already planned to some extent, the use of hydrogen as an alternative fuel cannot occur on a large scale until the next generation but one of aircraft, ie from around 2010 at the earliest. But it also means that it is necessary to carry out the relevant research and development work in good time, and the magnitude of the task dictates that this can be achieved only on an international scale.

The first trials with positive results were carried out in 1956/57 by NASA with a converted twin-jet B 57 Canberra, while Lockheed at that time was engaged in investigations including engine tests for supersonic aircraft.

Even in the subsonic range liquid hydrogen makes it possible to reduce the starting weight of high-capacity aircraft by half, compared with hydrocarbon-fuelled aircraft of the same range and payload. Because of the smaller total weight it is possible to achieve shorter take-off runs, steeper climbing paths (less nuisance from noise) and/or smaller engine thrust (less noise production) (Figure 14).

It may also be possible to decrease the size and weight of the engines (Table 2).

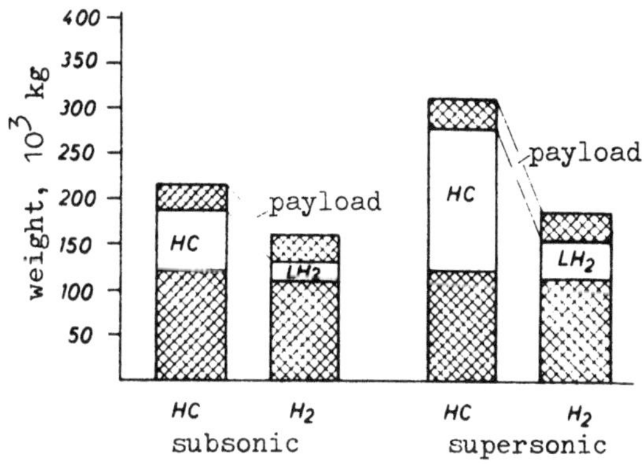

Figure 14 Comparison of the weight distribution in subsonic and supersonic aircraft with hydrocarbons or pure hydrogen as fuel

Table 2 Comparison between LH_2 and Jet A as fuels in high-capacity aircraft

		LH_2	Jet A	Ratio (Jet A/H_2)
Total weight	kg	178 000	236 200	1.33
Payload	kg	44 000	44 400	1.0
Empty weight	kg	110 000	111 800	1.01
Weight of fuel	kg	24 000	80 000	3.24
Thrust per engine	N	127 700	145 400	1.14
Wing area	M^2	313	389	1.24
Wing span	M	53	59.2	1.12
Length of fuselage	M	56.7	60	0.90
Take-off distance	M	1900	2435	1.28
A/W (cruising altitude)	-	16.07	17.91	1.12
Specific fuel consumption	kg/Nhr	0.203	0.590	2.92
Cost	$ 10^6	26.9	26.5	1
Energy consumed	kWh/seat km	0.43	0.48	1.12
Noise level:				
lateral	E PN dB	87.2	87.8	-
overflight	E PN dB	89.2	94.2	-

Table 3 Capital investment for the production of 1500 t of liquid H_2 per day (corresponding to 4000 t of Jet A per day as, for example, at San Francisco Airport in 1971). The output of electricity required for the production of hydrogen by electrolysis and for liquefaction amounts to about 4.5 GW

		Capital costs
Electrical energy (DM 1500/kW)	4.5 GW	DM $6.75 \cdot 10^9$
Electrolysis plant (DM 100/kg H_2/day)	$1.5 \cdot 10^6$ kg/d	DM $0.15 \cdot 10^9$
Liquefaction plant (DM 200/kg H_2/day)	$1.5 \cdot 10^6$ kg/d	DM $0.3 \cdot 10^9$
Tanks (DM 1000/m^3)	25 000 m^3	DM $0.025 \cdot 10^9$
Capital investment		approx. DM 7000 million

25 years' amortization
80% utilization
6% interest

LH_2	approx. DM 100/MWh
Jet A (at present)	DM 30-40/MWh
Motor vehicle fuel, including tax	DM 100/MWh

Hydrogen's favourable diffusion properties and high thermal conductivity lead to better mixing than with liquid hydrocarbons, even with shorter combustion chambers. The wide range of ignition for hydrogen-air mixtures (5% to 75% by volume) makes the engines more readily controllable, especially under partial loads, and reduces the emission of noxious substances. Since it is possible to use engines of shorter construction, thus reducing the time for which the fuel is at a high temperature, the emission of nitrogen oxides (in any case small with hydrogen) can be still further reduced. The problems, in addition to that of the economic production of hydrogen, lie in the provision of the necessary infrastructure at international airports and the bridging of the transitional period, which will necessarily be a long one (Table 3). A further problem concerns the placing of the fuel tanks, which are more bulky than those for kerosene. It is not possible to site them in the wings, since the surface/volume ratio of cryogenic containers must be kept as low as possible in order to keep down evaporation losses. Foam insulation can be used, in accordance with current technical practice, but the optimum construction requires further investigation. With present techniques double-walled tanks with an intervening vacuum for insulation would be too heavy. External tanks located on the wings would increase the aerodynamic drag considerably. Placing the tanks in the extended fuselage represents a reasonable compromise between space utilization, aerodynamic drag and safety. The insertion of suitable bulkheads would provide adequate safety even in case of accidents (Figure 15).

The heat required to vaporize and heat up the hydrogen for the engines can be obtained through certain sections of the outer skin of the wings and fuselage. In this way the boundary layer is cooled so as to produce laminar flow, resulting in a lowering of the aerodynamic drag and hence of the fuel consumption (laminar flow control - LFC): this could not be achieved to the same extent in any other way.

In addition to hydrogen some attention has been paid to liquid methane as a possible alternative fuel. Apart from the necessity for cryogenic storage, methane is more closely allied to hydrocarbon-based fuels. For the same amount of stored energy it requires twice as much volume as kerosene, so that in this respect it lies somewhere between hydrogen and conventional hydrocarbons. It can be produced by the gasification of coal, which raises the same environmental problems as the production of conventional synthetic fuels. Methane gas mixes more readily with the combustion air in engines than do conventional fuels, and thus causes less emission of noxious substances.

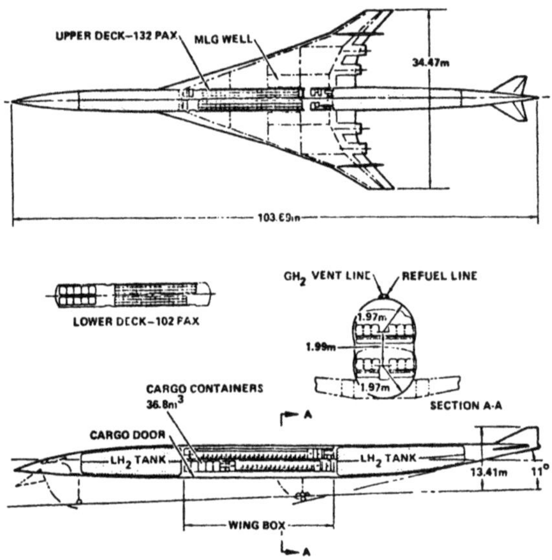

Figure 15a Possible layout for LH_2 tanks in the fuselage of a supersonic passenger aircraft

More serious problems arise in connection with safety in case of accidents, and these questions need further investigation both for methane and for liquid hydrogen. Experiments on the bursting of tanks have shown that liquid hydrogen evaporates and disperses extremely rapidly. These experiments included the bursting of a tank containing several thousand litres of liquid hydrogen, which evaporated in a few minutes. At temperatures above 33 K gaseous hydrogen at atmospheric pressure is lighter than air, and is dispersed correspondingly rapidly. This requires a temperature rise of only 13 K above the boiling point (20 K), while for liquid methane a temperature rise from 105 K (the boiling point) to 213 K, ie by more than 100 K, is necessary. The danger of spreading and fire at ground level is thus similar for liquid methane and the conventional liquid fuels, while hydrogen behaves more favourably in this respect. Furthermore, fires involving hydrogen last only one tenth as long as kerosene fires evolving the same amount of energy, so that the radiation from the hydrogen flames is only one tenth as great as that from kerosene flames. It is also important to note that no explosions or detonations have so far been observed in experiments carried out in the open air or in accidents involving hydrogen tankers. In spite of these findings it remains an important task to investigate the safety aspects further under conditions as close as possible to those obtaining in practice. However, present experience gives no grounds for supposing that hydrogen is more dangerous than hydrocarbons (Figure 16).

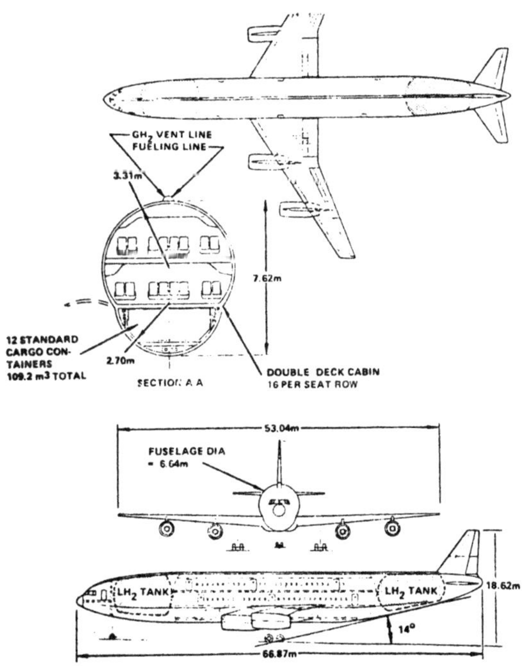

Figure 15b Possible layout for LH_2 tanks in the fuselage of a subsonic passenger aircraft

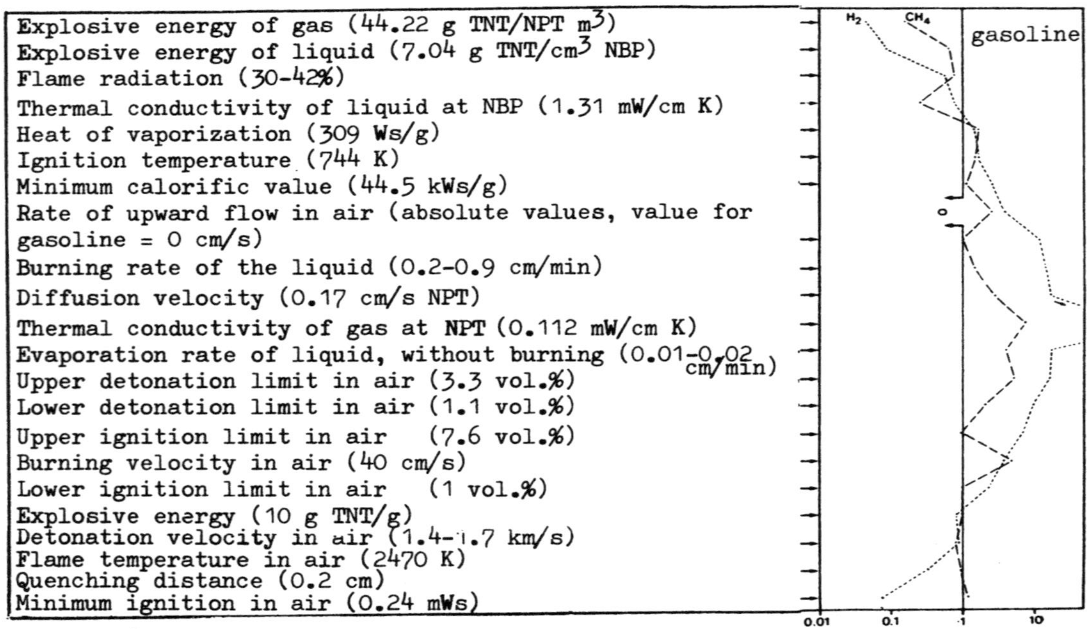

<u>Figure 16</u> Comparison of the properties which determine safety for hydrogen, methane and gasoline, referred to those for gasoline. The values in parentheses refer to gasoline

Hydrogen as an alternative fuel for motor vehicles

The consequences of using hydrogen in motor vehicles are best seen by comparing it with other chemical energy vectors. Table 4 shows the required tank volume and weight of fuel for typical non-conventional fuels compared with iso-octane, which is typical of the various hydrocarbons that constitute gasoline. Ammonia and hydrazine are included for comparison only out of interest. Ammonia will in practice burn only in pure oxygen and, like hydrazine, is excluded in any case on account of its toxicity. Comparison of iso-octane with methane (liquid natural gas), methanol and hydrogen shows that, although liquid hydrogen requires a larger tank volume, it requires the lowest weight of fuel on account of its high minimum calorific value. Except for the tank volume, liquid methane corresponds most closely with gasoline, while methanol requires an appreciably larger tank volume and also (on account of its lower calorific value) a considerably heavier weight of fuel (Table 5).

In the use of hydrogen as an alternative fuel, the storage of the fuel on board the vehicle constitutes an important problem. For motor vehicles hydrogen could in principle be stored as compressed gas, chemically as a metal hydride, or as a liquid. The storage of compressed gas would be difficult on grounds of safety because of the high pressures required. Furthermore, even at pressures around 200 bar the quantity of hydrogen stored amounts to only about 1% of the weight of the container, while the volume of the container is only 20 times greater than that of an equivalent gasoline tank. In metal hydrides hydrogen atoms are incorporated in the crystal lattice of metals or alloys (intermetallic compounds), and each lattice atom can absorb up to two hydrogen atoms, or in individual cases between two and three. When using alloys like iron-titanium the storage material can contain only up to 1.75 wt.% of hydrogen, since in iron titanium hydride ($FeTiH_{1.95}$) the iron and titanium constitute about 98.25 wt.% of the substance. The amounts of hydrogen stored are more favourable when the lattice consists of lighter atoms, for example magnesium.

Table 4 Comparison of various alternative fuels and their storage

	Storage capacity (wt.%)	Density of hydrogen (g/l)	Energy density kWh/kg	kWh/l	Operating temperature (at 2 bar)	α^2
Magnesium hydride	7	101	2.33	3.36	350° C	0.32
Magnesium nickel hydride	3.16	81	1.05	2.69	250° C	0.27
Iron titanium hydride	1.75	96	0.58	3.18	-10° C	0.14
Liquid hydrogen	100	71	33.3	2.36	-250° C	-
Methanol			5.6	4.42		
Liquid methane			13.8	5.8	-161° C	
Ammonia			5.14	4.21		
Hydrazine			4.6	4.65		
Iso-octane[1]			12.7	8.76		

1. Taken here as representative of gasoline, which is a mixture of low-boiling hydrocarbons.

2. Represents the ratio of the heat of formation of the hydride to the minimum calorific value of the hydrogen. This amount of heat must be supplied to the storage unit during operation in order to liberate the hydrogen.

Table 5 Weight required for the storage of fuels

	Hydrogen (kg)	Hydride (kg)	Storage container, etc (kg)	Total weight (kg)
Magnesium hydride	8.5	121		193
Magnesium nickel hydride	8.5	269		430
Iron titanium hydride	8.5	485		775
Liquid hydrogen	8.5	-	50	58.5
Equivalent quantity of gasoline	22.3	-	15	37
Equivalent quantity of methanol	50.5	-	20	70

Weight required to store 8.5 kg of hydrogen: comparison between metal hydrides, liquid hydrogen (120 l), gasoline and methanol.
The storage volume is about the same for metal hydrides and liquid hydrogen. It is about 32.5 l for gasoline and about 64 l for methanol. According to (10) the total weight of the hydride storage unit can be estimated by multiplying the weight of the hydride itself by 1.6.

However, other problems then arise, since the formation of the hydride and its decomposition to give hydrogen occur rapidly enough to produce a sufficiently high hydrogen pressure only at fairly high temperatures (250-350°C, see Table 1).

This temperature can be lowered by suitable additions to the alloy, particularly of nickel. However, this causes the heat of formation of the hydride, which must be removed when the storage unit is being charged and supplied when the hydrogen is being liberated, to assume very high values (about 30% of the minimum calorific value of hydrogen, see Table 4). This can cause problems in running the vehicle, since on account of the high operating temperature of these hydrides the heat available from the exhaust gases is insufficient to liberate from the hydride the hydrogen required by the engine. Since hydride vehicles have already been investigated intensively, a considerable amount of information is now available about the performance and potential of mobile hydride storage units.

For a given amount of energy stored in the vehicle, hydride storage requires about the same volume as liquid hydrogen, but 10-20 times the weight. Hydride vehicles must therefore be regarded as alternatives to electric vehicles, and would be used mainly for urban transport within a moderate range. However, it is precisely in urban transport that carrying the additional weight of the storage system is a disadvantage; for example, it may amount to about 300 kg for a maximum range of 150 km. The consequent increase in fuel consumption in urban transport makes the advantages that hydride storage offers over storage as liquid hydrogen, such as the possibility of loss-free storage at ambient temperatures and the absence of energy consumption for the liquefaction of hydrogen, seem less attractive. For the European driving cycle, the extra consumption of energy in the form of hydrogen resulting from the weight of the storage unit corresponds roughly to the extra energy consumption for liquefying hydrogen (cf Table 6). Thus, although the consumption of primary energy for liquefaction may be regarded as a particular disadvantage, it does not represent a limitation in comparison with hydride storage, since on the one hand liquid hydrogen and gasoline are comparable as regards weight and range even when the storage tank is taken into account, and on the other hand a considerable proportion of the energy used in liquefaction can be regained as mechanical energy when the cold hydrogen is warmed up. It can also be seen from Figure 17 that from the energy point of view liquid hydrogen corresponds roughly to synthetic hydrocarbons (synthetic gasoline from coal). It is also of importance for engine operation that a pump for liquid hydrogen can be used to obtain the extra pressure needed for internal mixing in the spark-ignition engine or indeed for diesel operation, with only a small expenditure of energy.

Comparison of hydride storage and storage of liquid hydrogen (LH_2) with regard to the additional amount of primary energy consumed in storage for the SAE 227 driving cycle[7]

Range 150 km	Weight (kg)[1] m_{H_2}	m_{MH}	m_{SP}	Primary energy in kWh/kg H_2 preparation	driving cycle	filling storage unit	Total kWh/kgH_2
$FeTiH_{1.95}$	2.84	162	260	2.6[2]	22.75[4]	4.8[5]	30.15
LH_2	2.33	-	15	31.4[3]	1.85[4]	0.05[5]	33.3
Range 500 km							
$FeTiH_{1.95}$[1]	21.6	1233	1974	2.6[2]	76.65[4]	16.00[5]	94.25
$FeTiH_{1.95}$[6]	13.6	835	1000	2.6[2]	56.76[4]	16.00[5]	75.36
LH_2	8.1	-	50	31.4[3]	5.96[4]	0.17[5]	37.53

1. m_{H_2} refers to the hydrogen to be stored, m_{MH} to the metal hydride and m_{SP} to the weight of the storage unit ($m_{SP} = 1.6\ m_{MH}$)
2. Primary energy required to compress gaseous hydrogen to the charging pressure of the storage unit (30 bar)
3. Consumption of primary energy, including liquefaction. The following values have been used: mechanical work for liquefaction 11 kWh/kg H_2, power station efficiency 0.35, electrolysis efficiency 0.8
4. Efficiency of engine, including transmission $\eta_M = 0.17$
5. Based on weekly tank filling over 8 years (384 fillings)
6. $m_{SP} = 1.2\ m_{MH}$
7. Apart from consumption when idling, the energy requirements of the EEC cycle correspond to those of the SAE-227 cycle employed here (see (26)). The figures given here therefore apply also to the EEC cycle

Table 6 Additional consumption of primary energy for the storage of hydrogen, based on the amount of hydrogen conveyed

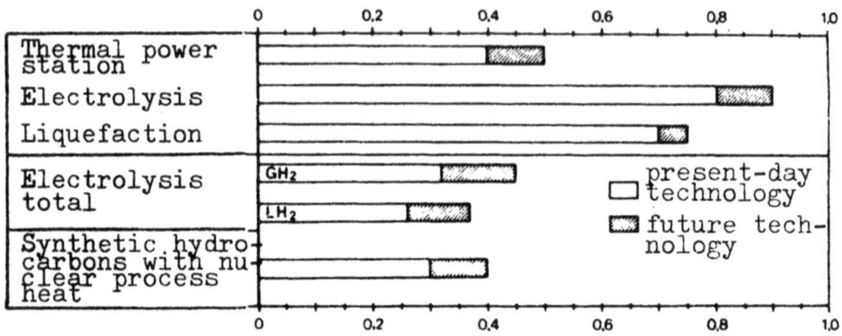

Figure 17 Expenditure of primary energy in H_2 liquefaction compared with that required for synthetic hydrocarbons

This would be difficult to achieve with hydride storage: either hydrogen at ambient temperatures would have to be compressed with a compressor, which because of the large volume of gaseous hydrogen would require an appreciable proportion of the output of the vehicle engine, or the injection pressure of the hydrogen into the engine would have to be reduced so much that the fuel could no longer be supplied at the most favourable moment.

A problem could arise with evaporation of hydrogen gas from the storage tank of the vehicle, especially when it is parked in an enclosed space such as a garage. However, according to the most recent observations the rate of evaporation from the latest vacuum-insulated storage containers can be kept down to a very low level by taking suitable precautions (Figure 18). Laboratory storage containers are now available for liquid hydrogen with capacities of 100-200 litres which can be kept for up to one year before all the liquid has evaporated. Storage tanks for motor vehicles giving storage periods of weeks or months are technically quite feasible today at reasonable cost. The small amount of hydrogen which evaporates can be used to generate electricity in a fuel cell, thus making it possible to dispense with a rotary electric generator. In this way the car battery can be charged efficiently and with little consumption of fuel when the car is stationary

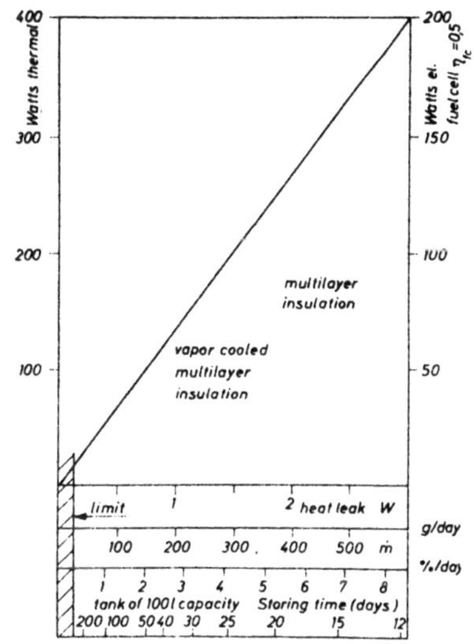

Figure 18 Evaporation from small storage tanks for liquid hydrogen. Storage times of more than 100 days (corresponding to an evaporation rate of less than 2% per day from a 100-l container) can be readily attained today. With such a container, if the energy in the evaporating hydrogen is converted into electricity with 50% efficiency in a hydrogen-air fuel cell, this will provide a steady available electrical output of about 20 watts, which can be used to charge the batteries of a stationary vehicle

and during town driving, and the emission of hydrogen to the surroundings, which is particularly dangerous in enclosed spaces, is avoided completely (Figure 19).

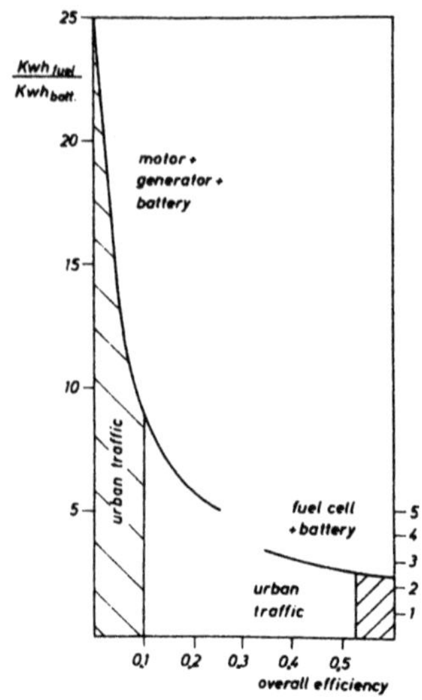

Figure 19 Consumption of primary energy (fuel) for battery charging in urban traffic. Comparison between conventional practice and the fuel cell

Infrastructure and fuelling

The main problems in the possible introduction of an alternative fuel lie in the creation of an adequate infrastructure, a network of filling stations and a system for organizing supplies to these filling stations. The impossibility of using the existing gasoline filling stations for hydrogen, together with questions of its economic production, constitute the chief arguments against the introduction of hydrogen in road transport in the medium term.

At first sight these problems appear to be simpler in the case of metal hydride storage, since here hydrogen can be handled merely at ambient temperatures. However, in order to fill the storage unit of a vehicle in a reasonable time, charging pressures of up to about 30 bar must be used, which means that the storage unit of the vehicle must also be designed for such pressures. Furthermore, during the charging of the storage unit the considerable heat of formation of the hydride (14% of the minimum calorific value for iron titanium hydride and 28% for magnesium nickel hydride) has to be removed by additional cooling, which requires the removal of considerable amounts of heat when several vehicles are being fuelled simultaneously.

When liquid hydrogen is used fuelling can be carried out by the method commonly used for gasoline, as has already been demonstrated. Figure 20 shows a computer-controlled filling appliance for motor vehicles, which is comparable with the self-service dispensing pumps in common use today in the ease with which it can be used. Faulty operation, which could lead to the overflow of liquid hydrogen, is virtually precluded by the computerized control. The time required to fill a 120-litre tank is at present between 5 and 8 minutes, and could be further reduced by the improvement of a number of components. It should be possible to realize a fuelling time of about 3 minutes.

The organization of supplies for a hydride filling station requires the provision of hydrogen gas under pressure, while liquid hydrogen filling stations can be supplied by tank trucks in the same way as for gasoline. Tank trucks for liquid hydrogen were developed in the USA specifically for the space travel programme, and are in regular use there on the normal road network.

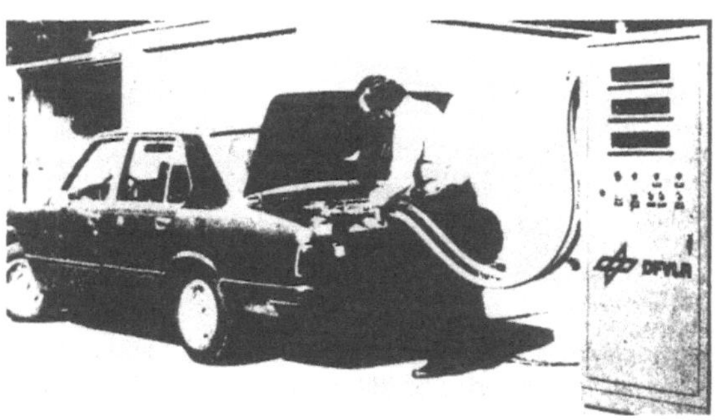

Figure 20 Semi-automatic microprocessor-controlled fuelling appliance for liquid hydrogen

Figure 21 Experimental vehicle with liquid hydrogen. This motor vehicle is the first in Europe to be fuelled with stored LH_2

Figure 22 LH_2 motor vehicle and fuelling arrangements: a joint demonstration project between the Los Alamos Scientific Laboratory and the DFVLR

By analogy with present-day refineries the liquefaction of hydrogen can be carried out conveniently in large central installations, either where the hydrogen is produced or at centralized sites with a good hydrogen supply.

Safety

Safety aspects are of primary importance to acceptance of the large-scale introduction of any secondary energy vector. In this respect hydrogen must be compared with other gaseous and liquid energy vectors, especially with gaseous and liquid methane and with low-boiling liquid hydrocarbons. Widespread public use demands the clarification of safety in transport, storage and handling, as well as the analysis of possible accidents, including their control and prevention.

All three secondary energy vectors mentioned here as examples have different characteristic properties, consideration of which does not, taken overall, suggest the preferential use of any one of them. All three have different fields of application, involve different risks and require a detailed investigation of safety in each individual case.

Hydrogen is used on a large scale today in industry, especially the chemical industry, where the safe handling of hydrogen is an established technique. Nor are any essential problems to be expected in this area of use in the future. However, in considering the possible future public use of hydrogen as an energy vector, for example as an alternative fuel, it has to be borne in mind that at present hydrogen (and especially liquid hydrogen) is generally considered an extremely dangerous substance to handle, although completely realistic assessments can be made on the basis of extensive experience in space travel. This is because such experience is not sufficiently widely known, and because the public has been accustomed for nearly a hundred years to dealing with gasoline and kerosene, although these too are by no means free from danger. Natural gas is also generally accepted today as an energy vector, although in the 1940s it gave rise to a number of extremely serious accidents with liquid gas tankers and large-scale storage in the USA. In the case of hydrogen this "familiarization effect" has not yet developed, although in the past hydrogen has been used in considerable quantities in town gas, which contains between 40% and 70% of hydrogen.

Although safety problems with hydrogen are somewhat different, in the present state of knowledge no greater potential dangers are to be envisaged than in dealing with natural gas, while the unequalled environmental compatibility of hydrogen must be counted an additional advantage.

On this account efforts to instil acceptance of hydrogen will need to include concrete applications within the framework of suitable demonstration programmes, in which it can be shown that no unreasonable or insurmountable safety restrictions need be imposed.

Summary

In comparison with other important secondary energy vectors, such as electricity, natural gas, synthesis gas or remote heating, hydrogen satisfies a number of the necessary conditions for a potential future secondary energy vector. These include availability, manufacture without harm to the environment, good storage properties and very ready conversion into other forms of secondary energy.

A coupling of hydrogen and electricity in a future energy economy is technically feasible. Hydrogen can thus successively replace petroleum-based products, including natural gas, and also synthesis gas.

Hydrogen can also be mixed with natural gas or synthesis gas, or can be used in a pure state: it can therefore play an important part in ensuring gas supplies when fossil energy vectors become scarce.

Hydrogen can play a specially significant role when

coupled with regenerative primary energy sources such as the sun and wind: since it can be stored, it can serve here to even out a fluctuating supply of energy and to transport energy to areas of high demand.

Hydrogen is also a candidate among the various alternative transport fuels such as synthetic hydrocarbons, methanol and liquid natural gas. Because of its favourable environmental characteristics it can be used both in air transport and in surface transport.

In view of the requisite order of magnitude of a future hydrogen market, however, enormous capital investment will be required for production, distribution and storage, and this factor represents the essential obstacle to a rapid introduction of hydrogen.

References

(1) J. Williams (Ed): Carbon Dioxide, Climate and Society, Pergamon Press, Oxford (1978)
(2) W. Bach and W.H. Matthews: Exploring Alternative Energy Strategies, Energy, Vol 4, pp 711-722, Pergamon Press, 1979
(3) G.M. Woodwell: The Carbon Dioxide Question, Sci.Am.January 1978, pp 34-44
(4) R. Dahlberg: Die Ablösung der fossilen Brennstoffe durch Wasserstoff (The replacement of fossil fuels by hydrogen)
(5) J. Gretz: On the Potential of Solar Energy Conversion into Hydrogen and/or other Fuels, Int.Journal of Hydrogen Energy, Vol 5, No 3, June 1980, pp 269-281
(6) Deutscher Bundestag Bericht der Enquete - Kommission "Zukünftige Kernenergie-Politik" (Deutscher Bundestag Report of the Commission of Inquiry on "Future Nuclear Energy Policy"), Deutscher Bundestag, Drucksache 8/4341, Bonn, June 1980
(7) C. Carpetis and W. Peschka: A Study on hydrogen storage by use of Cryoadsorbents, Int. Journal of Hydrogen Energy, Vol 5, No 5, pp 535-555
(8) C. Carpetis: A System Consideration of Alterna-

tive Hydrogen Storage Facilities for Estimation of Storage Costs, Int. Journal of Hydrogen Energy, Vol 5, No 4, pp 423-439

(9) P.O. Carden and L. Paterson: Physical and Energy Aspects of Underground Hydrogen Storage, Proc. 3rd World Hydrogen Energy Conference, Tokyo 1980, pp 797-811 (Vol 1)

(10) H. Buchner: The Hydrogen Hydride Energy Concept, Proc. 2nd World Hydrogen Energy Conference, Zürich 1978 (Vol 4)

(11) J.J. Reilly and G.D. Sandrock: Hydrogen Storage in Metal Hydrides, Sci.Am, February 1980, Vol 242, No 2, pp 98-106

(12) DFVLR: Wasserstoff als Sekundärenergieträger. Vorschlag für ein Forschungs- und Entwicklungsprogramm (Hydrogen as a secondary energy vector. Proposal for a research and development programme), Stuttgart, December 1980, DFVLR-MiH, 81-10; 1981

(13) H. Wojkowsky, W. Schnurnberger, H.J. Sternfeld and W. Peschka: Stromerzeugung aus Wasserstoff (Generation of electricity from hydrogen), Proc. "Hydrogen as an Energy Vector" Status Seminar, Comm. of European Comm., 12-14 February 1980, Brussels

(14) C. Carpetis: Large-scale Hydrogen Storage by Use of Cryoadsorbents and Comparison to Alternatives, Proc. "Hydrogen as an Energy Vector" Status Seminar, Comm. of European Comm., 12-14 February 1980, Brussels

(15) K. Strasserm: Development of an H_2/O_2/Fuel-cell Assembly with Circulating Electrolyte in a Compact Modular Design, Proc. Meeting Electrochemical Society, Boston, May 1979

(16) R. Noyes: Application of the Alsthom/Exxon Alkaline Fuel-cell System to Utility Power Generation, Final Report EPRI EM-384, January 1977

(17) S.S. Borys and J.P. Ackerman: Molten Carbonate Fuel-cell Systems Development Program, Proc. 14th IECEC, August 1979, Boston

(18) R.D. Witcofski: Potentials and Problems of Hydrogen-fuelled Supersonic Aircraft, Proc. 10th Intersoc. Energy Conversion Engineering Conference (IECEC), 1975, Paper 729211

(19) G.D. Brewer: Characteristics of Liquid Hydrogen-fuelled Aircraft, Proc. Symp. on Hydrogen in Air Transportation (DFVLR-DGLR), Stuttgart, September 1979

(20) G.D. Brewer: LH_2/Airport Requirements Study, NASA CR-2700, NASA - Langley Res. Center, 1976

(21) BMFT: Neuen Kraftstoffen auf der Spur (On the trail of new fuels), Study, Bonn 1974 (including 503 literature references)

(22) W. Stewart and F.J. Edelskuty: Logistics, Economics and Safety of a Liquid Hydrogen System for Automotive Transportation, Intersociety Conference on Transportation, Denver, Colorado (Los Alamos Rep. La-UR 74-1637)

(23) W. Peschka and C. Carpetis: Cryogenic Hydrogen Storage and Refuelling for Automobiles, Int.Journal of Hydrogen Energy, Vol 5, No 6, pp 619-627

(24) T.R. Strobridge: Cryogenic Refrigerators, An Updated Survey, Nat.Bureau Standards Techn. Note 655; 1974

(25) W. Peschka: Operating Characteristics of an LH_2-fuelled Car and of a Semi-automatic Refuelling Station, Proc. 3rd World Hydrogen Energy Conference, Tokyo 1980

(26) W. Peschka: Wasserstoff als Alternativkraftstoff im Kraftfahrzeug (Hydrogen as an alternative fuel in motor vehicles), Internationales Verkehrswesen No 6, November/December 1980, pp 447-454

(27) C. Carpetis: Comparison of the Expenses required for the Onboard Storage Systems of Hydrogen-powered Vehicles, to appear in 1981 in Int. J. of Hydrogen Energy

(28) W. Peschka, F.J. Edeskuty and W.F. Stewart: Liquid Hydrogen Storage and Refuelling for Automotive Applications, Proc. 3rd Miami Int.Conf. on Alternative Energy Sources, 15-18 December 1980, Miami Beach, Florida; ed. Int. Association of Hydrogen Energy

(29) J. Hord: Is Hydrogen Safe? National Bureau of Standards, NBS Technical Note 690

(30) F.J. Edeskuty, R. Reider and K.D. Williamson Jr: "Safety" in Cryogenic Fundamentals, Ed. G.G. Haselden, Academic Press, London/New York, 1971

Liquefaction of hydrogen in a large industrial plant

by H. Gutowski and M. Wanner[1]

1. Introduction

At the present time more than 90% of world energy requirements are met from primary energy sources of fossil origin, namely natural gas, petroleum and coal. The remaining 10% comes from nuclear reactors and hydroelectric plants.

Table 1 shows the percentage distribution between the different energy vectors in the European Community.

The progressive exhaustion of fossil primary energy sources and their resultant growing costliness demand far-sighted planning for the use of alternative primary energy sources and energy vectors. One of the energy vectors of the future is hydrogen, on account of its unlimited availability as a constituent of water, its combustion without harm to the environment, the fact that it can be stored, and the ease with which it can be transported as a liquid. However, its production and liquefaction require considerable amounts of energy, which can be justified economically only when energy is available at a low price and far removed from industrial centres with high energy demands.

Specially promising are plans to collect solar energy in deserts or on artificial ocean platforms (2,3) and to use it on a large industrial scale to decompose water into hydrogen and oxygen. Similar considerations apply to the use of hydropower in remote areas.

---

1. Dipl.Ing. H. Gutowski and Dr M. Wanner, Linde Aktiengesellschaft, Werksgruppe TVT München, Höllriegelskreuth

Table 1 (1)

Petroleum	53.3%
Gas	17.5%
Coal	22.3%
Hydro	3.7%
Nuclear	3.2%

In order to transport hydrogen to the consumers and in its further utilization, for example as a transport fuel, it is economic to liquefy it. The following low-temperature process is proposed for the liquefaction of hydrogen on a large industrial scale.

2. LH_2 plant with a capacity of 11 600 l/h of 98% para-hydrogen

2.1 Description of process (Figure 1)

The plant is shown in a simplified form, and is characterized 1) by the H_2-liquefaction line, in which the feed hydrogen is finally converted into the liquid product, and 2) by the H_2-circuit, which provides the necessary cooling by means of compressors followed by expansion in turbines.

The feed hydrogen is obtained at 0.21 kg/s by electrolysis at atmospheric pressure, and is first compressed to a working pressure of approx. 25 bar in order to facilitate the removal of heat from the liquid product by the H_2-circuit. After compression, but before introduction into the low-temperature process, the hydrogen passes through several stages of purification, such as the removal of traces of alkali, the catalytic conversion of O_2 to H_2O, and subsequent drying.

The feed hydrogen now has a purity of approx. 99.9%, and contains 0.1% of residual N_2. It is cooled to about 85 K in the countercurrent heat exchangers E1, E2 and E3. At this temperature the residual N_2 is removed by adsorption on silica gel.

During further cooling down to the liquefaction temperature of approx. 25.5 K the conversion of ortho- to para-hydrogen is effected catalytically in three temperature stages, namely at 60 K, 30 K and 25 K. Before coming into contact with the last catalyst the hydrogen, now cooled to approx. 29 K, is expanded to approx. 4 bar, when approx. 85% is liquefied.

The cooling required to condense the remaining 15% of gaseous hydrogen, to cool the whole liquid product further, and to remove the heat of the ortho-para conversion, is supplied by a liquefied branch stream of approx. 0.07 kg/s at approx. 3 bar and 24.5 K. The product itself, consisting of 98% para-hydrogen, is extracted at approx. 4 bar at approx. 25.5 K.

For the purpose of cooling, approx. 2.55 kg/s of H_2 is compressed from approx. 3 to 25 bar and then expanded in three turbines at different temperature levels. After cooling to approx. 134 K in the countercurrent heat exchanger E1 the total stream is divided. Approximately 97% passes through the turbines, while the remainder serves to cool the product to be liquefied in E7 and E8.

The first turbine T1 expands approx. 2.48 kg/s of H_2 in two stages from 25 to approx. 12 bar, and at the same time lowers the temperature from approx. 134 to 111 K: it produces a cooling effect of approx. 670 kW for an overall efficiency of approx. 72%.

After further cooling to approx. 85 K in countercurrent heat exchanger E3 further division takes place into 1.09 kg/s for turbine T2 and 1.39 kg/s for turbine T3.

The pressure of approx. 12 bar still available from the first turbine is brought down to approx. 3 bar in the two following turbines. This pressure drop is used up in two stages in T2, accompanied by a temperature drop from approx. 85 to 59 K, while T3, which has inlet and outlet temperatures of approx. 45 and 29 K respectively, requires only one stage for the same Δp.

The turbine outputs are approx. 280 kW and 180 kW for T2 and T3 respectively, corresponding to overall efficiencies of 74% and 76%.

The expanded circulating hydrogen is used for counter-current cooling in heat exchangers E6 to E1 of both the product stream to be liquefied and the branch stream split off after E1, which effects peak cooling down to 24.5 K.

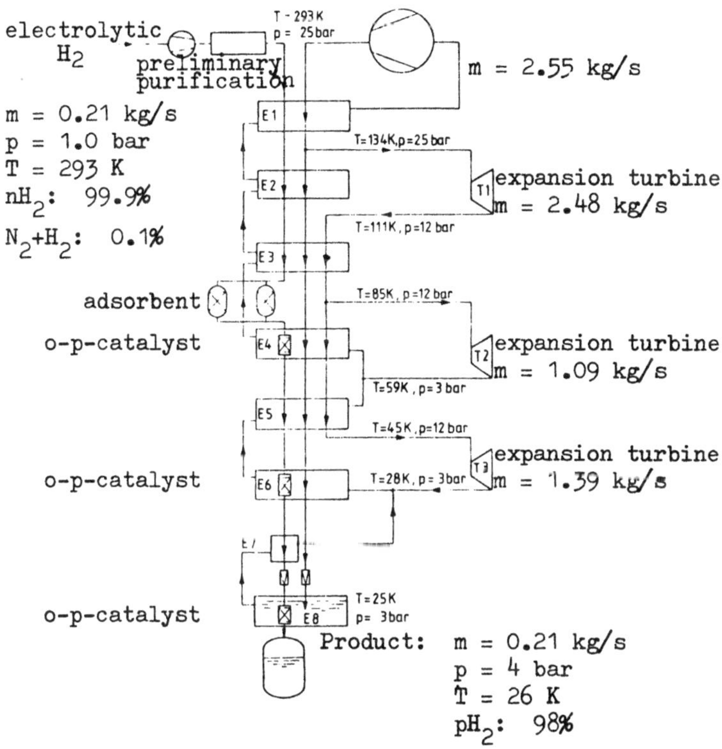

<u>Figure 1</u> Flow sheet for the H_2 liquefier

2.2 Details of the cooling circuit

2.2.1 Cooling effect required

The cooling required in this low-temperature plant is composed of the following contributions:

- Heat exchange losses

  For a 3.5° temperature difference at the warm end of countercurrent exchanger E1 these amount to approx. 150 kW

- Ortho-para conversion

  Starting with feed hydrogen at 293 K composed of 75% o-H_2 and 25% p-H_2, and compared with the LH_2 product at 26 K composed of 2% o-H_2 and 98% p-H_2, this contributes approx. 108 kW

- LH_2 product

  The difference between the heat content of the hydrogen gas supplied and the liquid hydrogen produced amounts to approx. 800 kW

- Insulation and adsorption losses

  Losses due to the penetration of heat into the cold box and to the regeneration of the N_2 adsorbent amount to approx. 72 kW

These contributions add up to a total cooling requirement of approx. 1130 kW

2.2.2 Refrigeration

The following limiting conditions were laid down for the refrigeration system:

- Production of LH_2 containing 98% para-H_2 at 26 K and 4 bar.
- No precooling with liquid N_2: this renders the plant more independent, with a view to locations with cheap energy supplies from hydropower.

- The use of H_2-turbines at three temperature levels, so as to ensure a favourable energy consumption.
- To improve reliability and reduce costs, the number of turbine stages used to effect the required pressure drop at one temperature should not exceed two.

The following parameters were selected in order to meet these conditions, and also to take into account the fact that the energy consumption per kg of LH_2 is significantly lowered by making the suction pressure of the circuit compressor as high as possible (1):

The final and suction pressures of the H_2-circuit compressor are 25 and 3 bar respectively. The latter value is determined by the pressure of the LH_2 product, ie 4 bar, given that the magnitude of the Joule-Thomson stream has to be kept low.

The final pressure of 25 bar arises from the arrangement of the turbines. The first of these, T1, expands the whole amount (apart from the 3% for the Joule-Thomson stream) in two stages to an intermediate pressure of 12 bar, while the second and third, which are in series with T1 and in parallel with one another, each expand about 50% of the stream to the final pressure of 3 bar. T1 works in the temperature range from 134 to 111 K, T2 in two stages from 85 to 59 K, and T3 in a single stage from 45 to 28 K.

2.2.3 T-S diagram for n-hydrogen (Figure 2)

The T-S diagram illustrates the thermodynamic cycle of the process for cooling hydrogen. It shows particularly clearly the dependence of the enthalpy difference bridged by the turbine expansion on the pressure conditions in the three temperature ranges. These values are particularly significant, since they determine the cooling effect produced by the turbines and the compressor energy used for this purpose. The smaller the angle between the turbine expansion and the isobars and the closer to 90° the angle between the turbine expansion and the lines of constant enthalpy, the greater will be the enthalpy difference bridged, and

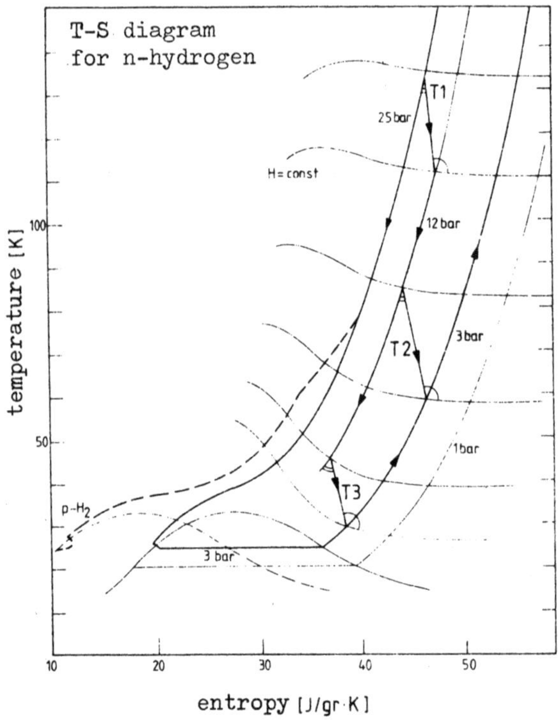

<u>Figure 2</u> T-S diagram for n-hydrogen

hence the cooling effect produced. Figure 2 shows that these conditions are best fulfilled for the first turbine, and become increasingly worse at lower temperatures for T2 and T3.

Table 2 shows for the three turbines the pressure drop produced, the cooling effect Q_T, the compression energy Q_K required per kg of H_2, and the ratio $\xi = Q_T/Q_K$.

Table 2

	P_E/P_A	Q_T $\frac{kJ}{kg}$	Q_K $\frac{kJ}{kg}$ isothermal	Q_K $\frac{kJ}{kg}$ compressor	$\xi = \frac{Q_T}{Q_K}$ isothermal	$\xi = \frac{Q_T}{Q_K}$ compressor
T_1	2.08	268	892	1500	0.3	0.179
T_2	4.0	296	1690	2800	0.175	0.106
T_3	4.0	134	1690	2800	0.08	0.048

2.2.4 Thermodynamic efficiency - exergy

An estimate of the thermodynamic efficiency of the refrigeration process can be made with the help of the exergy, ie the ability to do work.

According to Baehr (4) the exergy of a state is defined by

$$E_i = H_i - H_u - T_u (S_i - S_u) ,$$

where H_i and S_i are the enthalpy and entropy of the state in question, and H_u and S_u those at the temperature T_u of the surroundings.

The exergy of the hydrogen product is thus equal to the minimum work which would have to be done to carry out the liquefaction process reversibly. This assumes a continuous ortho-para conversion as the temperature changes. Because of the thermodynamic losses in the system the actual work required is greater than the exergy of the liquid hydrogen produced. These losses arise from irreversible changes in the heat exchangers, in the compressor and in the turbines, and from the expansion in the Joule-Thomson valves. In addition there is a loss because the heat of the ortho-para conversion is not absorbed at a continuously varying temperature but, as already stated, in three temperature stages.

Table 3 summarizes these exergy losses. Figure 3

shows graphically the flow of exergy and the percentage contribution of the individual components to the total loss. The losses in the heat exchangers come partly from the final temperature differences and partly from insulation losses. E4, E6 and E8 also contain the ortho-para conversion losses.

Table 3 Summary of exergy losses

Components	Loss of exergy kW
Circuit compressor	4 780
Compressor for feed gas	520
Turbines	1 840
Heat exchangers, o-p conversion	1 640
Throttles	10
Total losses	8 790
Exergy of liquid p-H_2	2 800
Output of turbines	1 130
Actual consumption of compressors	12 720

2.2.5 Thermodynamic losses in the H_2 liquefier (Figure 3)

Comparison shows that the hydrogen compressors make the largest contribution to the irreversibility of the whole process. The turbines and the heat exchangers represent roughly equal sources of loss, while the loss of exergy due to the throttles is vanishingly small.

The greatest possibilities for improving the process therefore lie in the compressors. Since their efficiencies are already typical for heavy-duty compressors, it would appear difficult to achieve any substantial improvements in this sector.

The losses in the turbines could be reduced to a limited extent by improvements in efficiency.

If the energy liberated in the turbines is not utilized

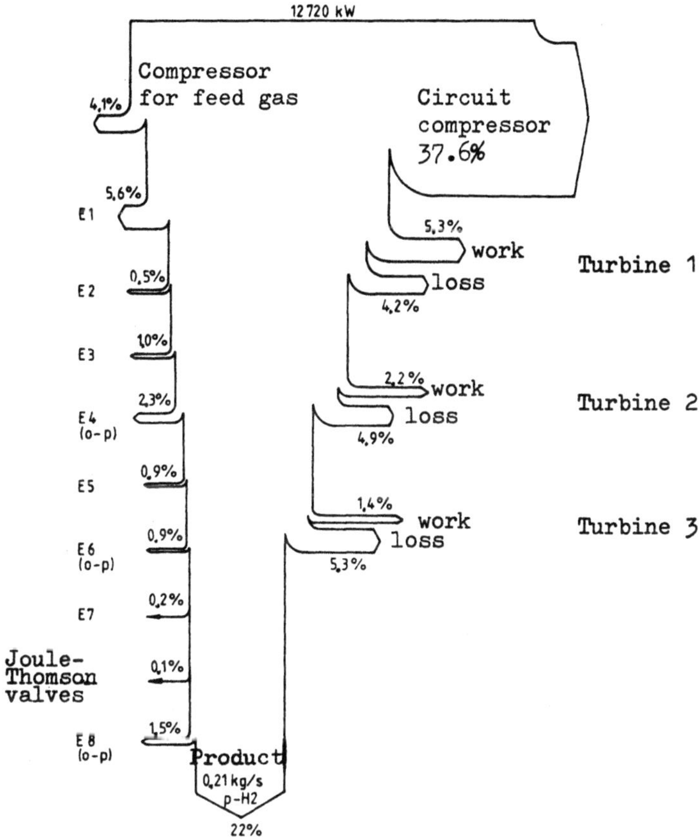

Figure 3 Exergy flow sheet for the H_2 liquefier

this hydrogen liquefier attains a thermodynamic efficiency of 22%. This corresponds to a specific energy consumption of 60.6 kJ/g of liquefied para-hydrogen.

If the work yielded by the turbines is utilized the efficiency of the whole installation can be further improved. Assuming that 80% of the output of the turbines can be fed back electrically, a thermodynamic efficiency of 23.7% and a specific energy consumption of 56.3 kJ/g can be obtained.

2.3 Some equipment and engineering features

The insulation of the equipment is provided for by a cold box filled with slag wool. With a view to its transport as a unit package it has a base measuring approx. 4 x 4 m and a height of approx. 14 m.

The H_2-turbines employed are immersed in oil, equipped with generators, and are also constructed as unit packages. The dimensions of a turbine cold box are: base approx. 3 x 5 m, height approx. 3 m.

These boxes are flushed out with hydrogen and are maintained at a slight excess pressure (approx. 20 mm head of water) in order to prevent the ingress of air.

The compression of the electrolytic hydrogen and in the H_2-circuit is effected by dry-running two-stage piston compressors. This type is preferred to rotary compressors on the grounds of energy saving, which is of the order of 5-10%. This advantage is not cancelled out by any increase in costs, as may be seen by comparing the cost of the four parallel piston compressors required with that of a single rotary compressor. An increase in the number of compressors also leads to more reliable operation, since if one of them fails the plant can continue to operate at a reduced load. The turbines are fitted with swivelling nozzles and can tolerate load fluctuations between 50 and 130%.

2.4 Consumption figures

Table 4 gives a summary of the most important consumption figures for the liquefaction of 0.21 kg/s, corresponding to 11 600 l/h of liquid hydrogen.

In addition to the main consumption by the hydrogen compressors, some electric power is required for running vacuum pumps, for regenerating the adsorbent and for control purposes. Taking into account the recovery of 80% of the output of the turbines, the total electricity consumption is barely 12 MW.

Cooling water is needed essentially for removing the heat produced in the hydrogen compressors. There is a small consumption of hydrogen arising from leaks in the compressors and turbines, and from the regeneration of the adsorbent and flushing out the cold boxes.

Table 4 Consumption figures for the H_2 liquefaction plant

Feed gas:		
electrolytic hydrogen 0.21 kg/s = 8500 m^3/h (n.t.p)		
99.9% pure, atmospheric pressure		
1. Consumption of electric power		
circuit compressor		11 380 kW
compressor for feed gas		1 340 kW
other electrical equipment (pumps, heating, control)		50 kW
recovered from turbines		- 900 kW
Total electric power		11 870 kW
2. Cooling water consumption	approx.	950 m^3/h
3. H_2 lost by leakage	approx.	$3.7 \cdot 10^{-3}$ kg/s = 150 m^3/h (ntp)

2.5 Installation and production costs

The capital costs of an installation ready for use are estimated at DM 40 million, and the specific energy requirements for the feed and circulating compressors amount to 15.6 kWh/kg of LH_2. Assuming an annual charge on capital of 17%, which is usual for this kind of plant, this gives an annual expenditure of DM 6.8 million.

If the cost of electricity produced by hydropower is DM 0.07/kWh and the annual running time is 8000 h, then the annual energy costs amount to DM 6.6 million.

Annual expenditure on maintenance and repairs is estimated at DM 1 million. Charges for cooling water and other materials will amount to DM 0.4 million per annum, and the same figure applies to personnel costs. The total annual expenditure on return on capital and running costs thus amounts to DM 15.2 million. Referred to an annual production of 6048 tonnes of LH_2, this gives a production cost of DM 2.50/kg. Conversion to energy, using the minimum calorific value of 30 000 kcal/kg for hydrogen, results in an energy price of DM 83.33/Gcal. Comparing this with other liquid energy vectors such as natural gas and petrol, for which the energy prices to large consumers are at present approx. DM 30/Gcal and DM 70/Gcal respectively, it is seen that there is no question of hydrogen being competitive at present: however, it is very likely to become so in the medium term.

Estimates carried out in 1977 (5) led to the conclusion that on economic grounds the realization of hydrogen-based energy could not be expected in the present century. These estimates assumed an annual increase of 6% in the cost of materials, interest rates and energy. However, the experience of the last few years has shown that these figures cannot be maintained, especially for energy, but must be revised upwards. This implies an acceleration in the introduction of the use of hydrogen for energy supplies. Examples of this are the testing of LH_2 as a propellant for aircraft in the USA and projects for the introduction of LH_2 for reducing ores in Canada.

References

(1) Jahrbuch für Bergbau, Energie, Mineralöl und Chemie (Yearbook for Mining, Energy, Oil and the Chemical Industry), 1980/81

(2) W.J.D. Escher: A Macro System for the Production of Storable, Transportable Energy from the Sun and the Sea, American Chemical Society, 163rd National Meeting, Boston, 1972

(3) E.J.D. Escher and J.A. Hanson: Ocean-based Solar-to-Hydrogen Energy Conversion Macro System, 1973

(4) H.D. Baehr: Thermodynamik (Thermodynamics), Springer-Verlag, 1973

(5) G. Kandler: Die Energieversorgung mit flüssigem Wasserstoff - Wirtschaftlichkeitsbetrachtungen durch Vergleich mit LNG-Import (Energy supplies with liquid hydrogen - Economic aspects by comparison with imported LNG), Gas, Wärme, International 26, No 8, August 1977

Utilization of nuclear process heat for the production of hydrogen

by H. Barnert[1]

Abstract

The possible applications of nuclear process heat as a form of new primary energy vector for the production of secondary energy vectors, particularly hydrogen, are discussed. The high-temperature reactor as a source of high-temperature process heat is described and details are given of the state of research and development in this sector. A brief report is given of the R&D work in the "Nuclear process heat prototype plant" and "Nuclear secondary energy" projects. The use of nuclear process heat for the production of hydrogen and oxygen from water is discussed and the state of R&D work in the field of thermochemical cyclic processes is described.

1. Introduction

In addition to the use of nuclear energy for electricity generation, it is planned to use thermal energy from nuclear reactors, particularly in the form of process heat, for the production of new energy vectors which can serve to replace oil and natural gas. As a result, nuclear energy will also come to be useful for heating purposes over the long term. It is apparent that the combination of nuclear energy and coal will play a special part.

Of the developed nuclear systems, because of its technical design the high-temperature reactor (HTR) in

---

1. Dr Heiko Barnert, Institut für Reaktorentwicklung, Kernforschungsanlage Jülich GmbH, Jülich 1

particular, as an advanced reactor system, has the capability of making available a number of technologies based on the integration of process heat into an energy conversion process. It is of particular importance here for the Federal Republic of Germany (1,2) that indigenous coal resources can be utilized, by means of coal beneficiation processes, to cover the demand for non-polluting and convenient energy vectors.

Research and development work has been going on for more than a decade in Germany and other countries on the use of nuclear energy in the form of high-temperature heat as process heat. As long ago as 1968 the use of high-temperature heat for coal beneficiation was suggested (3) and an extensive development programme accordingly initiated. The present objectives of research and development work in this sector are, naturally, those which can be achieved in a relatively short time, eg coal beneficiation using nuclear process steam of superheated steam quality, as well as such longer-term possible applications as the production of hydrogen and oxygen from water using high-temperature heat. Proposals for these were discussed at international level as long ago as 1969 (4).

2. The HTR as a source of process heat

Taking a general view, nuclear process heat represents a new form of primary energy source (5) which can be used in many different ways for the production of secondary energy vectors for the heating market. In these conversion processes, hydrogen generally plays an important part, both as an energy vector and as a chemical raw material or a reducing agent for metal production.

The development of the HTR builds on the experience gained from the operation of some 40 gas-graphite reactors which so far have produced about one quarter of the electricity generated in the Western World. The characteristic feature of the HTR is the supply of heat at a high temperature level. So far five HTRs have been commissioned or are under construction in the world: Dragon in the UK, Peach Bottom and Fort St.

Vrain in the USA and AVR and THTR-300 in the Federal Republic of Germany. The Dragon and Peach Bottom reactors have now gone out of service according to schedule; the THTR-300 is at present under construction.

The significance of the two German reactors for the use of nuclear process heat will now be explained briefly.

The experimental nuclear power plant in Jülich belonging to the Arbeitsgemeinschaft Versuchsreaktor (AVR) GmbH has in its 13 years of operation (6) provided scientific, technical and economic information and experience which is of great importance for current development work. The AVR reactor is a pebble-bed HTR which has been operated since 1974 with an increased mean gas exit temperature of 950°C - the only reactor in the world to do this. Thus it has been demonstrated that the HTR is suitable for the production of high-temperature process heat.

The very encouraging experience obtained with the AVR in power production and in experiments includes, in particular, the high level of availability in terms of time and energy, the production of heat at a high temperature, the demonstration of safety characteristics, fault-free operation of the blowers and other installations, and also the low radiation dose for the staff and the environment. Because of a leak in the steam generator, AVR had to shut the reactor down in May 1978, but operation resumed in August 1979. The variety of experience obtained during this incident (7) is of great value for future development. The mean gas exit temperature of 950°C demonstrated by the AVR reactor is the reference temperature for two process heat applications, namely methane reforming in a tube furnace and steam gasification of coal via intermediate heat exchangers. It is also a standard value for other process heat applications, such as the production of hydrogen and oxygen from water using HTR heat (nuclear splitting of water).

The THTR-300 prototype nuclear power station (8), belonging to Hochtemperatur-Kernkraftwerk GmbH, Uentrop bei Hamm, has been under construction since 1972. A major difficulty with the construction of the THTR-300 has been the prolongation of the construction period, which is now predicted to be 140 months (9). The power station is at present about 70% complete (10): all the main buildings are up, the installation of the main turbo-set is complete, the prestressed concrete pressure vessel is finished, the six steam generator units are installed, the 1 million fuel elements have been mass produced and the blowers are about to be fitted.

The main reasons for the delay in construction have been the extensive re-designing and the further verification which became necessary because of changed requirements in the licensing procedure.

3. Survey of the various processes

At present, the nature of standard manufacturing processes means that hydrogen is usually mixed with other high-energy gases, eg carbon monoxide and methane. Only for a few applications in the chemical industry is it necessary to produce pure hydrogen. For energy purposes, on the other hand, it may be useful to consider gas mixtures and other gaseous hydrocarbons or even liquid hydrocarbons as well as pure hydrogen. The conversion of hydrogen and carbon monoxide into methane and methanol has been described in this connection as the addition of hydrogen atoms to carbon atoms for the purpose of improving transportability and storability.

Hydrogen can be produced substantially from water, hydrocarbons and coal (the latter contains some hydrogen); generally, energy is needed for this and it can be obtained from coal, hydrocarbons and nuclear energy.

From the point of view of the application of nuclear energy in these conversion processes, the proportion of nuclear energy used should be as high as possible.

The following classes of process, described below, are given in order of technological difficulty and therefore of possible date of introduction:

1) process steam + coal + (H_2O) : H_2

2) process heat + coal* + (H_2O) : H_2

*or: hydrocarbons (natural gas, petroleum)

3) process heat + (H_2O) : H_2, O_2

3.1 Nuclear process steam for Lurgi pressure gasification

There have been recent suggestions (11) for using HTR process steam for the beneficiation of hard coal by the Lurgi pressure gasification process, with the established technology and in particular with the "Ruhr 100" advanced technology (12). The product is a gas rich in hydrogen which can be treated in further conversion stages to become pure hydrogen.

The use of nuclear energy for this conversion process involves the provision of process steam of superheated quality (for "Ruhr 100": 100 bar, 400°C) and of electricity for the process, particularly the production of oxygen from air.

In this conversion process the heat-producing component is the steam generator. This technology can therefore be implemented without any further fundamental R&D work.

Against the background of the continued interest in the HTR for electricity generation and with a view to the gradual introduction of nuclear coal gasification, in the summer of 1980 an electricity supply company group, comprising Ruhrkohle AG and Ruhrgas AG headed by the Preussische Elektrizitäts AG, set up an operator group to be jointly responsible for the design and later construction of an HTR for producing electricity and process steam.

3.2 Process heat for coal beneficiation or the conversion of hydrocarbons

The basic alternatives for the introduction of process

heat in the form of high-temperature heat into a coal beneficiation process are "hydrogasification" and "steam gasification of coal", which are being developed in the "Nuclear process heat prototype plant (PNP)" project (13).

In the "hydrogasification of coal" the nuclear process heat is introduced via the tube furnace (methane reforming) and the steam generator. The reforming of methane in the tube furnace produces the hydrogen needed for hydrogenation. The products of the process can be substitute natural gas (SNG) or hydrogen.

In the "steam gasification of coal" the nuclear process heat is introduced as high-temperature heat via an intermediate circuit directly into the coal gasifier and via a steam generator. The nuclear heat in this case is used for heating the gasification reaction. The primary product is a synthesis gas which for various applications can be converted into SNG or hydrogen; see also (14).

In the PNP project, work has been in progress since 1975 on the development of processes for the conversion of solid fossil raw materials with heat from high-temperature reactors, the purpose being, after completion of the design work, to build and operate a prototype plant with a thermal capacity of eg 500 MW (15), and then to try for large-scale industrial use. The results of the work done so far for the nuclear heat generating system of the PNP-500 prototype plant can be described as follows. The reactor core can be designed for a helium gas exit temperature of 950°C, the shutdown equipment can be designed for a limited operating time, the primary heating gas piping can be made and the coupling of different processes is considered feasible. There are problems with present knowledge of the lifetime of the lateral reflector and the shutdown equipment. The feasibility of extrapolating the heat exchange components, ie the tube furnace and intermediate heat exchanger, to high numbers of operating hours has yet to be verified. Further efforts need to be made to reduce the tritium contamination of the product gas below the manufacturing tolerance; the latest results on coating heat transfer surfaces have indicated a possible process. By making the whole reactor containment building inert or partly inert, reactions between the atmosphere in the building with any escaping process gas can be reliably prevented.

It may be of particular importance for a future hydrogen industry that the "nuclear reforming" process offers the possibility of producing hydrogen from natural gas and petroleum fractions using high-temperature heat. The basic process is being developed in the "Nuclear secondary energy (NFE)" project (16). The raw materials are a hydrocarbon, eg methane, and steam and the primary product is a mixture of hydrogen, carbon monoxide, carbon dioxide and methane, the hydrogen being predominant. As part of the NFE project there is a semi-industrial plant and a pilot plant for the reforming and methanation reactions (known as EVA and ADAM) in operation at Kernforschungsanlage Jülich.

The "Nuclear secondary energy (NFE)" system can be used as a "half-open system" to supply synthesis gas or hydrogen as well as secondary energy (17).

On the subject of coal beneficiation processes using nuclear heat, it should be further noted that the use of process steam from a pressurized water reactor has also been proposed for coal gasification (18). In this case the process steam is taken in the form of slightly superheated steam from the live steam line behind the steam generator.

3.3 Process heat for the production of hydrogen from water

The most sophisticated technology for the conversion of nuclear energy in the form of high-temperature heat into a non-polluting and convenient secondary energy vector is the production of hydrogen and oxygen from water by means of the thermochemical cyclic process. A thermochemical cyclic process is one for converting thermal energy into chemical energy, eg in the form of hydrogen and oxygen from water. Thermodynamically, this conversion process is a sub-class of the generalized thermodynamic cyclic process for the conversion of heat into secondary energy vectors (19), and it is therefore comparable with the thermomechanical cyclic process for the production of mechanical energy or electricity.

R&D work in the field of thermochemical cyclic processes has been done in many institutes and agencies throughout the world for about a decade, eg (20). The state of R&D work can be described as follows. In many institutes work is still being done on the development of new cyclic processes. Of the many cyclic processes so far put forward, the family of "sulphuric acid cycles" is distinguished by the fact that series of tests are being conducted on it on a semi-industrial scale. In particular, development work is at a highly advanced stage at the Ispra Joint Research Centre (21).

The heat-input equipment of thermochemical cyclic processes is comparable in principle with the process equipment for introducing heat used hitherto or under development. The tube furnace is an example of an endothermic high-temperature reaction with gaseous reaction partners, and the coal gasifier using steam is an example of such a reaction between solids and gases. R&D work is being done along these lines in the PNP and NFE projects, which is also of great significance for the development of thermochemical cyclic processes.

The drive to develop and investigate new thermochemical cyclic processes is the result, in particular, of general realization that materials technology is a problem in the development of the cyclic processes of the sulphuric acid family. Proposals for "hydrocarbon cyclic processes" have been justified from this point of view, using methane reforming, for example, as heat-input reaction (22); there are, of course, problems here in closing the cycle process on the low-temperature side.

In addition to the possibility of producing hydrogen and oxygen from water using nuclear heat, there is of course the possibility of water electrolysis. Since electricity is used in this process, it is in general an extension of the thermomechanical cyclic process. Electrolysis does not fall within the scope of this paper.

Compared with the electrolytic production of hydrogen and oxygen from water, the thermomechanical cyclic process can claim that it improves fundamentally the primary conversion of thermal energy as regards the efficiency of the conversion. For this reason this technology is more sophisticated and more fundamental. It is generally assumed today that if development work is successful, the industrial use of thermochemical cyclic processes can be expected after the turn of the century.

4. Summary

The use of nuclear process heat for the production of hydrogen, or in general for the production of secondary energy vectors for the heating market, is justified mainly by the fact that substitutes must be found for oil and natural gas. As an advanced reactor system, because of its technical properties and the provision of thermal energy at a high temperature level the high-temperature reactor can be used in particular to supply high-temperature process heat as well as for generating electricity. This process heat can be used for the beneficiation of fossil energy sources, such as coal, in the production of new secondary energy vectors for the heating market, ie hydrogen in particular.

References

(1) V. Hauff: Die Rolle des Hochtemperaturreaktors bei der Energieversorgung der Bundesrepublik Deutschland, HTR-Statusbericht, 26 September 1978 (The role of the high-temperature reactor in the energy supplies of the Federal Republic of Germany HTR Status Report, 26 September 1978), JÜL-Spez-21, November 1978, pp 9 and 12.
(2) BMFT: Bericht des Bundesministers für Forschung und Technologie über die Entwicklung des Hochtemperaturreaktors an den Ausschuss für Forschung und Technologie und an den Haushaltsausschuss des Deutschen Bundestages (BMFT: Report of the Federal Minister for Research and Technology on the development of the high-temperature reactor to the Committee on Research and Technology and the Budget Committee of the German Bundestag), 5 June 1978
(3) R. Schulten: Über die Anwendung von Hochtemperatur-Kernreaktoren zur Kohlevergasung (The use of high-temperature nuclear reactors for coal gasification), Arbeitsgemeinschaft für Forschung des Landes Nordrhein-Westfalen, No 185, 171st meeting on 12 June 1968 in Düsseldorf
(4) P. Capriopglio, C. Marchetti, R. Schulten, K. Knoche and G. Debeni: Round table on direct hydrogen production, Ispra, 12 December 1969
(5) R. Schulten and K. Kugeler: High-temperature reactor and application to nuclear process heat, Annals of nuclear energy, Vol 3, pp 95-112, 1976
(6) Arbeitsgemeinschaft Versuchsreaktor GmbH, Hochtemperatur-Reaktorbau: 10 Jahre Stromerzeugung mit dem Kugelhaufenreaktor der AVR (10 years' electricity generation with the AVR pebble-bed reactor), 17 December 1977
(7) K. Krüger and W. Dehring: Steam generator leakage in the AVR nuclear power station, ENC-79 Conference of the European Nuclear Society, 6-11 May 1979
(8) G. Schwarz and F. Schweiger: Stand der Weiterentwicklung der THTR-Linie in der BRD (State of further development of the THTR system in the FRG), and other papers, Atomwirtschaft, Vol 16, No 5, May 1971

(9) W. Mattick: Energiepolitische Aspekte (Energy policy aspects), BBC-Nachrichten, No 8/9, 1980
(10) H. Harder and M. Simon: Entwicklungsstand des Hochtemperaturreaktors (State of development of the high-temperature reactor), BBC-Nachrichten, No 10, 1980
(11) Kernforschungsanlage Jülich GmbH: Kohlevergasung mit HTR-Prozessdampf, Rahmenvorstellungen für eine stufenweise Einführung der nuklearen Kohlevergasung mit Hochtemperaturreaktoren (Coal gasification with HTR process steam, Outline proposals for the phased introduction of nuclear coal gasification with high-temperature reactors), 22 May 1979
(12) H.P. Peyser and G. Röbke: Die Versuchsanlage Ruhr 100 zur Weiterentwicklung der Lurgi-Druckvergasung von Steinkohle (The Ruhr 100 experimental plant for the further development of Lurgi pressure gasification of hard coal), Chemie-Technik, 1980, pp 401-404
(13) PNP: Statusbericht zum Ende der Konzeptphase vom 1.8.1975 - 30.11.1976: Ergebnisbericht der Planungs-, Forschungs- und Entwicklungsarbeiten (PNP: Status report on the end of the design phase from 1 August 1975 to 30 November 1976: Report on the outcome of planning, research and development work), January 1977. Partners in the PNP project: Bergbauforschung GmbH, Gesellschaft für Hochtemperatur-Reaktortechnik mbH, Hochtemperatur-Reaktorbau GmbH, Kernforschungsanlage Jülich GmbH and Rheinische Braunkohlenwerke AG.
(14) K.H. van Heek: Kohlevergasung zur Produktion von Wasserstoff (Coal gasification for the production of hydrogen), see elsewhere in this publication
(15) H.J. Neef and I. Weisbrodt: Coal gasification with heat from high-temperature reactors, Nucl. Engineering & Design, 54 (1979), 157-174
(16) R. Harth: Nukleare Fernenergie, Jahresbericht 1975 der KFA Jülich (Nuclear secondary energy, Annual report 1975 of KFA Jülich). Partners in the Nuclear secondary energy (NFE) project: Rheinische Braunkohlenwerke AG and Kernforschungsanlage Jülich GmbH
(17) M. Walbeck and Scharff (Project leaders): Plan-

studie, halfoffene Fernenergieversorgung - Kurzfassung (Design study, semi-open secondary energy supply - Summary). Report of the Kernforschungsanlage Jülich, March 1980
(18) Kraftwerk Union AG Erlangen: Kohleveredlung (Coal beneficiation), Discussion paper, May 1979
(19) H. Barnert: Anmerkungen zur thermochemischen Produktion von Wasserstoff aus Wasser mittels Hochtemperatur-Reaktorwärme (Notes on the thermochemical production of hydrogen from water by means of high-temperature reactor heat), JÜL-1660, June 1980
(20) International Energy Agency (IEA): Proceedings of the 3rd Technical Workshop, Task I: Thermochemical processes, 18-21 June 1980, National Chemical Laboratory for Industry, Tsukuba Science City, Ibaraki, Japan, 1980
(21) G. Schütz: Thermochemische Prozesse zur Wasserstoffgewinnung (Thermochemical processes for hydrogen production), see elsewhere in this publication

Safety aspects in the use of hydrogen in industry and the home

by W. Balthasar[1]

1. Introduction

The hazards - both real and supposed - presented by hydrogen must be evaluated before it can be used as a fuel and energy vector in any particular system. The public's questions about safety will have to be answered if hydrogen is to be introduced for use as a transport fuel, for heating in the home and in industry, for new industrial processes such as generating electricity at periods of peak demand or, to a greater extent than present, for industrial processes such as ammonia, fertilizer and methanol production and coal gasification or liquefaction. This enlarged hydrogen market will be accompanied by a corresponding growth in the production of hydrogen, in new techniques for its production, in storage capacity and in transportation across or beneath public territory. All these elements of a "hydrogen energy system" are involved in any discussion of safety.

It must be stressed that hydrogen has been used as a fuel and as a chemical for more than two centuries. Paracelsus is attributed the discovery of hydrogen in 1500. At the beginning of the 18th century, Lemory showed that a mixture of hydrogen and air explodes in the presence of a flame. In 1781 Cavendish demonstrated that hydrogen burns in air to form water alone. This brief look back at the history of hydrogen shows that at that time hydrogen was discovered because of its ability to burn and to explode. It also indicates that hydrogen should be used only with appropriate safety precautions.

1. Dr Wolff Balthasar, Kinetics Technology International BV, Zoetermeer, Netherlands

Total world production of hydrogen in 1973 was 2.1×10^{11} m^3, which is equivalent to 1.8×10^7 tonnes (1). These figures show that considerable experience exists in the handling of hydrogen. Hydrogen is used mainly as an intermediate product in the chemical and petroleum industries, the three largest consumers being the production of ammonia and methanol and the refining of crude oil. In 1974 the combined demand from these three consumers accounted for 92% of all the hydrogen consumed in the USA (2).

Almost all the hydrogen in the petrochemical industry is produced and consumed in one place. Only a few small pipeline systems carrying hydrogen between several plants are in use. The largest hydrogen pipeline system is in Germany and has a total length of approx. 210 km and an annual throughput of approx. 3×10^8 m^3 (3).

A number of analytical reports on the safety of hydrogen have already been published (1,4,5,6).

2. Accidents with hydrogen

In a recently published report to the US Department of Energy, Zalosh and Short (7) summarized and analysed hydrogen accident reports. They used seven independent categories of causes to describe accidents occurring after an escape of hydrogen:

Group	% of accidents
unnoticed leakage	22
hydrogen/oxygen waste-gas explosion	17
pipe and pressurized-tank explosion	14
inadequate flushing with inert gas	8
ventilation-system accidents	7
hydrogen/chlorine accidents	7
other causes	25
Total	100

We have already described and analysed these categories in detail in an earlier report (8).

Another recent publication comes from Williams (9) and concerns the causes of shut-downs in ammonia plant. This publication is valuable because of the details given of the causes that led to the shut-down of plant wholly or partly engaged in producing hydrogen. The analysis can therefore be used to translate practical experience and derive conclusions. Furthermore, for the first time the questionnaire that was used asked about all fires that had occurred during the period covered, and not just fires resulting in shut-downs.

Of the 29 plants questioned, only two reported that they had had no fires. The remaining 27 had had one or more fires.

In all, 125 fires were reported as having occurred over a four-year period (1973-1976). Averaged over the 29 plants, this means that one fire occurred per plant every 11 months. The causes are given in Table 1. It must be stressed that fires originating from pipes, collectors and connections within primary reformer furnaces have not been included in these statistics.

Leaks from seals at flanges and valves are responsible for 65% of the fires. With the exception of two, all the fires recorded under the heading of seals and valve seals were caused by hydrogen-rich gas leakage. Fittings that resulted in the most leaks that in turn led to fires were located in heat exchangers upstream or downstream of the high-temperature methanation conversion process, in waste-heat boilers and in ammonia converters. Leaks at pipe flanges and valve flanges were distributed throughout the entire plant, but flanges at catalyst pipes above or below primary reformers were responsible for many of the flange leaks. All leaks at valve flanges that led to fires were in the synthesis circuit.

The three cooling-tower fires reported were all caused by pipe leaks in one of the coolers downstream of the synthesis-gas compressor, causing synthesis gas rich in hydrogen to be passed to the cooling towers.

Table 1 Plant components as causes of fires

Group		Number
Seals		46
Apparatus flanges	23	
Pipe flanges	16	
Valve flanges	7	
Valve seals		10
Oil leakage		24
Transfer collector		9
Auxiliary boiler		8
Primary reformer		7
Cooling tower		3
Electrical		2
Other		16
	Total	125

One fire every 11 months in an ammonia plant is a high incidence. However, it must be emphasized that this figure includes only those faults that actually resulted in a fire. Bearing in mind that most ammonia plants are sited in the open air with natural ventilation, it has to be concluded that many more leaks must have occurred - along with the attendant risk of fire and explosion - which were rectified by plant and maintenance personnel.

It is likewise apparent that a large discrepancy exists between the 342 faults in plant components in 30 large ammonia plants within two years and 125 fires in 29 ammonia plants within a period of four years compared with the 117 fires out of a total of 409 faults for a period of 12 years, as reported in the Zalosh study. It therefore has to be concluded that in fact many more fires and explosions occurred than are reported in the literature. At the same time it also has to be assumed that none of these fires or explosions led to injury or death or to extensive damage to plant, because if they had they would have been reported in the literature or included in statistics compiled by insurance companies or official inspectorates.

3. Failure of fittings and components in ammonia plant

As already shown above (Table 1), leaks at flanged connections and valve seals account for 65% of all the fires recorded in 29 ammonia plants between 1973 and 1976 (9). Of these fires, 23 were associated with flanges to tanks, 16 with flanges to pipes and 7 with valve seals.

The same finding can be deduced from the Zalosh study, in which 409 faults were examined: 81 of these faults were due to pipe failure (including leakage), 41 were due to valves or flanges and 52 were due to other seals.

This analysis of causes indicates the characteristic tendency of hydrogen to penetrate normally "airtight" seals. The volume that leaks through a seal is determined either by the density of the gas (molecular weight) or by its viscosity, for a given pressure difference and leakage area. Like methane, hydrogen has a lower viscosity and lower molecular weight than air; hence, a fitting or component that is normally airtight is by no means necessarily "gas-tight" to hydrogen or methane.

When leaking hydrogen has a temperature near or above room temperature it rises upwards rapidly owing to the difference in density. The density of hydrogen is only 1/8 that of methane; hydrogen therefore has a correspondingly faster rate of rising. Hydrogen's relatively high diffusion coefficient also promotes dispersion even under still atmospheric conditions.

In the event of an escape of hydrogen in an enclosed space, its tendency to rise imposes a limit on the degree to which it is diluted by the air. As will be recommended below, for this reason hydrogen should not be handled in enclosed spaces but always in the open air.

Waste-gas systems and furnaces (heat-treatment furnaces and reformer furnaces) are more often affected by faults than any other components. Compressors and heat exchangers are likewise frequently affected. The relatively high accident rate for compressors, furnaces, heat exchangers and reformers is understandable when it is remembered that these usually contain hydrogen at temperatures close to its spontaneous ignition point (585°C).

The relatively high accident rate for systems in which the operating personnel are not specifically familiar with the hazards presented by hydrogen should also be mentioned. This is especially noticeable in, for example, nuclear power stations, chlorine plants and semiconductor factories, where hydrogen is only an unwanted side product that has to be handled. Leaks from flanges and the failure of waste-gas collection systems - and in the case of chlorine, the gas collection system - are the fittings and components mostly affected by faults. In view of the possible use of water electrolysis as a method of producing hydrogen in the future, it is important to point out that electrolysis has a high accident rate, most of the accidents today occurring in chlorine/alkali plants. It is therefore particularly important to ensure that the hydrogen and the oxygen are kept separate in the electrolysis cell and in the gas collection system.

Traditional fittings and components like pressurized tanks, distillation columns and heat exchangers as such appear to be affected to a negligible extent by hydrogen faults.

4. Preventive measures

In discussing preventive measures it is necessary to distinguish between measures intended to prevent accidents inside the system and outside the system.

Three conditions must be fulfilled simultaneously for a fault inside a system: for example, hydrogen as fuel, oxygen/chlorine as oxidizing agent and, lastly, a source of ignition. The elimination of sources of ignition may be an adequate preventive measure for avoiding dust explosions, for example, but it is usually not adequate for avoiding hydrogen explosions because of the extremely small amount of energy required to ignite hydrogen. Consequently, to avoid explosions in hydrogen systems it is absolutely essential to prevent oxygen from entering the system. This can be accomplished either by making the pressure in the system higher than that of the oxidizing atmosphere or by using an inert gas within the system. It should be pointed out that only 4 vol.% of hydrogen is needed to cause an explosion provided conditions are suitable in other respects. Since the lower explosion limit of hydrogen is also dependent on pressure, particular attention has to be given to determining the lower ignition point for the particular gas composition and pressure concerned.

The inside of the system can be rendered inert not only by using nitrogen but also by using carbon dioxide, halons or extinguishing powders such as ammonium phosphate or potassium chloride.

The latter can be used for temporarily rendering mixtures of hydrogen and air inert in order to prevent explosions progressing through the system. In this case they are therefore actually explosion suppressors.

Another way of preventing explosions from propagating through the system is to install flame traps. Here too, reliable flame traps for hydrogen are difficult to design because hydrogen requires a very narrow gap to prevent a flame from passing.

Whilst a high internal pressure and an inert internal atmosphere efficiently and reliably prevent the formation of explosions in pure hydrogen systems, other measures are needed in systems containing hydrogen and other products to prevent the dangerous consequences of explosions.

These measures are essentially as follows:

1. systems that withstand the force of an explosion,

2. systems that withstand the force of pressure surges,
3. explosion-relief systems,
4. explosion-suppression systems and, for pipeline systems alone,
5. flame traps, flame suppressors, explosion-relief devices and rapid-closing devices.

A detailed description can be found in the relevant literature (10).

Measures for preventing hydrogen explosions outside the system are greatly limited in effectiveness because an oxidizing agent, air, is always present and the presence of a source of ignition always has to be assumed even though it may have previously been demonstrated that numerous leaks, and therefore the presence of hydrogen in the free atmosphere, had occurred without an explosion and that the hydrogen had diffused away harmlessly. Consequently a fundamental preventive measure in the case of hydrogen plants is that they be built in the open air. Further, it is very important to detect any hydrogen leaks as early as possible. Because of hydrogen's low density, standard analysers like gas detectors can be used to only a limited extent and can be made effective only in open-air plant. The other preventive measures are largely secondary measures giving protection solely against further effects resulting from fires and explosions. Examples of such secondary measures are automatic or manual shut-down of the process, sprinklers, flooding, water spraying, extinguishing systems, halon systems, and explosion-relief devices for entire buildings.

All these systems have proved their effectiveness in the past. Automatic sprinklers, flooding systems and water spraying for cooling have a success ratio of 80%. This result seems to confirm the advantages of the qualitative recommendation made in published guidelines on explosion prevention. Powder extinguishing systems have a success ratio of 86% in extinguishing hydrogen fires. Shutting down the entire hydrogen plant is the most natural method of preventing a fire and has a success ratio of 81%. Explosion-relief devices have a success ratio of 90% and have contributed greatly to preventing damage to apparatus and buildings.

An essential basic feature of every plant, and especially a large hydrogen plant, is the layout of all the various sections of the plant with respect to one another. Safety must be considered right at the design stage of a plant and must be borne in mind continually throughout the subsequent stages of the plant's construction and commissioning and, finally, operation. Various companies have given these aspects special attention and have formalized appropriate checks. Two systems have recently become particularly well known and they will be described briefly here.

The Dow Chemical Company has introduced a graded system of formalized checks applicable to the various stages of a project. The first safety check is performed after the basic-design phase has been completed and is conducted by a team composed of safety experts, process engineers and plant personnel who have many years' experience in the technology concerned. Fixed specifications are laid down for the layout and structure of the plant which are intended to limit the maximum possible damage and to minimize the spatial separation between the sections of the plant whilst taking fire and explosion indices into account. This group of experts uses a precisely prescribed procedure for conferring with the plant's designers. Recommendations made by the safety testing group are binding. At the end of the detailed engineering stage the documents are once again examined on the basis of the same principles as before. Finally, before the plant is commissioned one more formalized check is made. This check now includes the personnel who will actually operate the new plant.

Only after this check has been concluded successfully can the plant be put into operation.

Imperial Chemical Industries in the UK have developed

a system of hazard and operability studies, known as "hazop analysis". This also consists of formalized checks, performed largely on completion of the process-design and detailed-design phases and based on check-words such as "none", "more", "smaller proportion of", "more than" and "other possibilities". The safety testing group is composed of basically the same categories of people as in the Dow example, but in this case the process engineers and especially the reliability engineers play a larger role. In this author's view hazop analyses are a refinement of the Dow testing system in terms of design checks.

It is obvious that the operating personnel have to be trained to use their plant. This training must also include safety aspects in detail. Today it is customary (and in Germany obligatory) for practically all plants that handle hydrogen to have a safety engineer in addition to the normal personnel. One of the safety engineer's essential duties is to ensure that personnel are kept properly instructed in the possible hazards and that only those personnel are employed who are qualified to do the work involved. Safety engineers will therefore also make it possible to organize the continuous training, instruction and refresher training of personnel: for all the technical regulations that may concern the safety of the plant itself, man is and will remain the main factor affecting that safety. The safety engineer and his expertise will therefore be crucial to the success of safety precautions in a hydrogen plant.

5. Concluding remarks

Accidents have occurred with hydrogen where it has been used either as a fuel or as a chemical reagent in a process. These accidents have occurred despite the fact that the properties of hydrogen are well known. One might ask whether the relevant safety regulations are inadequate. In a recent publication Tanner (11) has shown - in agreement with our own conviction - that as far as the legal side of the question is concerned this is clearly not the case. Our experience with regulations within our own company for various types of plant allow us to state that these regulations ensure adequate protection for people and plant. Accidents with hydrogen are no more serious as regards injury to people and damage to plant than accidents with inflammable liquids or natural gas. Of a total of 409 accidents listed in the Zalosh study, only 22 resulted in death and 101 in injury. Only 18 accidents led to damage in excess of $ 1 billion.

When we look at the other side of the question and compare hydrogen accidents with accidents involving inflammable liquids or natural gas, we find that the causes are similar and run more or less along the following lines:

- Although appropriate regulations existed, they were not followed.

  This is usually due to lack of knowledge of the hazards or to inadequate training.

- A safety analysis of the process in question is necessary and should take into account all possible and imaginable hazards and their combinations or sequences of faults. Safety analysis is the basis of pertinent planning.

- Materials and parts of apparatus tend to fail.

  There are, however, many ways of monitoring materials and parts. The number of monitoring instruments used is naturally a matter of money: the value of the parts being protected must be weighed against the investment cost of the instruments. Plant failure and especially unpredicted sequences of failures will occur if safety analysis has not been correctly applied.

- Working conditions must also be adequate.

  It is intolerable to imagine someone who has not been appropriately trained working where there is a risk of hydrogen leakage.

Whether it is hydrogen or another gas, we frequently find that the person who has failed to apply his knowledge is the main cause of an accident.

The results of our investigations can be summed up by stating that handling hydrogen correctly leads to the safe operation of plant. We can see only two basic areas in which further progress would be desirable and for which we therefore recommend further studies. These are as follows:

1. the development of better seals for use with hydrogen, and
2. the development of systems for detecting hydrogen in open-air plant; the idea is to develop gas warning devices similar to those already used successfully and reliably for inflammable gases having densities higher than air.

Summary

Determining the safety hazards presented by any given fuel requires a large amount of work that involves assessing the technical data of the fuel itself and also comparing it with other fuels. Any fire or explosion hazard must be carefully defined to enable the relative safety of any given fuel to be determined in every possible application. It may therefore happen that hydrogen is safer than conventional fuels in some applications but more dangerous in others. The object of this paper has been to examine the safety aspects of hydrogen systematically, and to do so by analysing accidents involving hydrogen and by looking at the fittings and components affected by these accidents. General regulations and precautions for the design of hydrogen plant and reactive fire-prevention systems are then considered on the basis of this approach. This form of presentation is intended to show that the general answer to the question of the safety of hydrogen is that hydrogen can be handled safely provided the relevant precautions are observed.

In the light of the results a number of recommendations are then given for further basic research projects that would improve safety in the handling of hydrogen.

References

(1) R.G. Zalosh and T.P. Short: Compilation and Analysis of Hydrogen Accidents Report, DOE Contract No EE-77-C-02-4442

(2) D.C. Adkins and R.J. Jaske: Hydrogen, a chapter from Mineral Facts and Problems, 1975 edition, Bureau of Mines Bulletin 667, 1975

(3) J.H. Kelley and E.A. Laumann: Hydrogen Tomorrow: Demand and Technology Requirements, Report of the NASA Hydrogen Energy Systems Technology Study, Jet Propulsion Laboratory, December 1975

(4) Hydrogen Safety Manual, NASA, N 75-72909, 1968

(5) J. Hord: Is Hydrogen Safe? National Bureau of Standards, PB-262 551, October 1976

(6) J. Hord: Is Hydrogen a Safe Fuel? International Journal of Hydrogen Energy, 3, 157 (1978)

(7) R.G. Zalosh and T.P. Short: Comparative Analysis of Hydrogen Fire and Explosion Incidents, Factory Mutual Research Corporation, March 1978

(8) J.P. Schödel: Hydrogen - a Safety Risk, Hydrogen as an Energy Vector, EC-EUR 6085, October 1978

(9) G.P. Williams: Chem.Eng.Progr. 74, 88 (1978)

(10) W. Bartknecht: Explosionen (Explosions), Springer Verlag, 1978

(11) W. Tanner: Sicherheitsaspekte einer künftigen Europäischen Wasserstoff-Technologie (Safety aspects of a future European hydrogen technology), Hydrogen as an Energy Vector, EC-EUR 6085, October 1978

(12) The Dow Chemical Co, Safety and Loss Prevention Guide, Midland Michigan, 1976

(13) H.G. Lawley, Chem.Eng.Progr. 70, 45 (1974)

MIX
Papier aus verantwortungsvollen Quellen
Paper from responsible sources
FSC® C105338

If you have any concerns about our products,
you can contact us on
ProductSafety@springernature.com

In case Publisher is established outside the EU,
the EU authorized representative is:
Springer Nature Customer Service Center GmbH
Europaplatz 3, 69115 Heidelberg, Germany

Printed by Libri Plureos GmbH
in Hamburg, Germany